Seed Production of Selected Horticultural Crops

(Vegetables, Spices, Plantation crops, Flowers and Ornamentals)

Dr. A B Sharangi is an Associate Professor of eminence in the discipline of Horticultural Sciences and has been in the profession of teaching, research and extension in the Department of Spices and Plantation Crops, Faculty of Horticulture, Bidhan Chandra Krishi Viswavidyalaya (Agricultural University), India for about sixteen years. He is associated with the process of coconut improvement leading to the release of a variety Kalpamitra from CPCRI. He has published about 45 research papers in peer-reviewed journals, 40 conference papers, 12 reputed books including the one from Nova Publishers (USA) and scores of popular scientific articles. Presently he is associated with about 25 international and national journals as regional editor, technical editor, editorial board member and reviewer. One of his paper has ranked among the top 25 articles in ScienceDirect.

Dr Sharangi has visited abroad extensively on academic mission and obtained several prestigious international awards *viz.*, ENDEAVOUR Post-doctoral Award-2010 (Australia), INSA-Royal Society of Edinburgh Visiting Scientist Fellowship (UK, 2011), FULBRIGHT-Nehru Visiting Lecturer Fellowship (USA, 2013) etc. He has delivered a couple of invited lectures in UK, USA, Australia, Thailand, Israel and Bangladesh on several aspects of herbs and spices. He is associated with a number of projects having academic and empirical implications. He is also having the membership of several international scientific societies and national academies including the New York Academy of Science (NYAS), Scientific Committee and Editorial Review Board on Medical and Biological Sciences, World Academy of Science, Engineering and Technology (WASET), National Academy of Biological Sciences (NABS), to name a few.

Seed Production
of
Selected Horticultural Crops
(Vegetables, Spices, Plantation crops, Flowers and Ornamentals)

DR. A B SHARANGI

2014

Regency Publications

A Division of

Astral International Pvt. Ltd.

New Delhi - 110 002

ISBN 9789351302018

Published by	:	**Regency Publications** A Division of **Astral International Pvt. Ltd.** – ISO 9001:2008 Certified Company – 4760-61/23, Ansari Road, Darya Ganj New Delhi-110 002 Ph. 011-43549197, 23278134 E-mail: info@astralint.com Website: www.astralint.com
Laser Typesetting	:	**Classic Computer Services**, Delhi - 110 035
Printed at	:	**Replika Press Pvt. Ltd.**

PRINTED IN INDIA

World Noni Research Foundation

12, Rajiv Gandhi Road, Perungudi, Chennai-600 096, India.

Phone : 91-44-2454 5401 - 04 Fax : 91-44-2454 5406 E-mail : mail@worldnoni.net Website : www.worldnoni.net

M: 094465 13017, 093817 48878
Email: kvptr@yahoo.com

Prof. K.V. Peter, Ph D
Former Vice-Chancellor KAU
Director, World Noni Research Foundation,
No.12,Srinivasa Nagar, 2nd Street,
World Wellness Open University Building,
Rajive Gandhi Road, Old Mahabalipuram Road,
Perungudi, Chennai-600096.

Chennai – 600 096
18, September 2013

Foreword

Robert Louis Stevenson said, "Don't judge each day by the harvest you reap but by the seeds that you plant". Seed is indeed the most vibrant, valuable and vital input in agriculture when our desired goal is (0 produce crop plants for meeting our demand for food. It is with us from the day when we started our eternal journey, from pre-historic times, to this modern era, always protecting us with (he warmth and compassion of a mother to his child.

The present book addresses the principles of horticultural seed production covering selected vegetables, spices, plantation crops, flowers and ornamentals coupled with the latest scientific research towards this direction. The whole book has been well designed with systematic chapterization and supplemented with a couple of annexures covering important information and updates related (0 seed production, quality, import, export, quarantine issues, etc.

I believe, this book by Dr A B Sharangi, will certainly meet the long-felt needs of a good book in this field and will serve as a text cum reference book for the students, teachers, scientists, researchers and all the persons related to this area in general I also congratulate the publisher Astral International Pvt. Ltd. New Delhi for the excellent printing-error free and pleasing.

(K V Peter)

Prologue

It is said *"Subeejam Shushereto Jaayaty Sapadayaty"* which means the good seed in a good field produces abundantly. Seed is the basic, vital and central input in agriculture and horticulture. Crop productivity is directly related to the genetic potential of the seed planted. It is estimated that the direct contribution of quality seed alone to the total production is about 15–20% depending upon the crop and it can be further raised up to 45% with efficient management of other inputs. Production of good seeds of horticultural crops is now gaining relevance not only for the wide range of diversity of the crops but also for the increased global reliance for these popular, attractive and promising products offering their time-tested crucial role for the economic and nutritional security around the world.

The book "Seed Production of Selected Horticultural Crops" deals with the scientific approach to seed production of some common but important horticultural crops ranging from vegetables, spices, plantation crops, flowers and ornamentals. Information on various aspects for the production of quality seed of each of the crop including general aspects and botany, environmental and isolation requirements, detailed cultural requirements as well as harvest and post harvest management has been compiled and elucidated.

The aim of this text book is to collate the traditional knowledge of horticultural seed production with that of the recent advances made in this direction throughout the globe in a comprehensive manner. It is hoped that the book will be quite useful for horticulturists, seed technologists, entrepreneurs, amateur gardeners, scientists, scholars, UG and PG students.

Dated: January 1, 2014   A.B. Sharangi
Kalyani

Contents

Chapter 1

General Introduction

Horticultural crops include fruits, vegetables, flowers and ornamentals, spices and plantation crops, medicinal and aromatic plants, forest plants, etc. All the crops are propagated either by sexual or asexual or by both the means. Usually perennial crops are propagated vegetatively, whereas those of annuals are through seeds. In perennial crops, however, seeds may be utilized to raise rootstocks for breeding programmes.

Botanically seed is a fertilized ovule. Practically it is a unit of propagation and is capable to transmit genetic traits from generation to generation. It consists of a tiny embryo in the resting stage provided with food and well protected by the hard seed covering. The embryo has one or more cotyledons, which serves usually as foliage leaves after seed germination.

Use of good quality seed is vital for getting successful crop production. Our ancient literatures *viz., Manu Samhita* also advocated about the necessity of using good quality seed. Seed being the most essential input of crop production comprises of a small part of the total cost of inputs involved. But, it can spoil the whole effort of successful cropping if poor quality of seed is used.

Supported by the release of a large number of crop varieties, the growth of the seed sector, predominated by the public seed companies has reached to an annual turnover of about Rs. 500 – 600 crores by late 80s (Dravid, 2011).

Regular supply of quality seed to the farmers for an assured production is undoubtedly a big challenge particularly for a country like India. The process of quality seed production is not that simple. Rather it demands skillful manipulation of some complex operations *viz.,* selection of good planting material (*e.g.*, breeder seed), maintenance of proper isolation distance, removal of off types and wild relatives, etc. at proper time in a proper way, observation of field standards, proper agro-techniques including plant protection, maintaining strict post harvest requirements etc. Presently the huge demand of seed in our country is partly met by public and private sectors through indigenous seed production and the rest through imports. But considering the overall demand of seed is in an increasing order, we must concentrate on producing our seed to such an extent that it would not only meet our present requirement but also open the way for export in the days to come. This necessitates our scientists, researchers, breeders, entrepreneurs as well as the farmers to go hand in hand and work sincerely towards a significant contribution for our country.

Chapter 2

Seed

Seed is a mature ovule which consists of an embryo and surrounded by a protective coat. It is not merely a particular plant part that can be used for multiplication but also any part of the plant used for propagation in true sense.

2.1 Classification of Seed

Seed may be classified into four distinct types as is used in the seed production programme. The use of seed of an appropriate class and from an approved source is necessary for raising the seed crop. Four classes of seeds, namely, breeder's, foundation, registered and certified seeds have been defined by the Association of Official Seed Certification Agencies (AOSCA):

1. Breeder or nucleus seed: The seed or any vegetatively propagated material which is directly controlled by the originating breeder or institution. It produces for the initial or recurring increase of foundation seed.

2. Foundation seed: The seed stock designated or distributed by an agricultural experiment station and supervised or approved by representatives of the station (*viz.*, NSC, SSC, etc.) which most nearly maintains specific genetic identity and purity is known as Foundation seed.

3. Registered seed: The progeny of the foundation seed which is produced by progressive farmers according to the technical advice and supervision provided by NSC or SSC and maintain satisfactory genetic identity and purity is known as Registered seed. This is usually approved and certified by a certifying agency.

4. Certified seed: The progeny of the foundation or registered seed which is produced by progressive farmers annually according to standard seed production practices and maintains satisfactory genetic identity and purity is known as Certified seed. This is also approved and certified by a certifying agency like SSC and is available for general distribution to farmers.

2.2 Qualities of good seed

For higher net monetary return per unit area, input and time, quality of seed should be undoubtedly very good and it is the responsibility of the producer that seeds are of assured genetic and physical purity. The followings are the requisite of a quality seed:

(a) Genetic factors:

   i. Higher yield (at least 10-15% than the existing local or used varieties).

  ii.   Pest and disease resistance

 iii. Adaptability to a wide range

 iv. Responsiveness to better growth situations

  v. Resistance to adverse environment *viz.*, drought

 vi. Acceptability of quality in the market and by the consumers

(b) Physical factors

   i. Freedom from other varieties of same crop or other crops

  ii. Freedom from weed seeds, other crop seeds, inert matters or any other materials which impair quality

 iii. Freedom from seed borne diseases

 iv. Good germinability

  v. Uniformity in size, colour and weight

The seed with qualities ideal in all respect is very rarely achieved. That is why the minimum quality standards are fixed in a National Seed Programme (and maximum in the cases of disease, weed, etc.). For standardization of seed testing procedures the ISTA was established on

10th July 1924 having about 60 countries as members with a representation on the seed testing attributes.

2.3 Environmental factors affecting vegetable seed production

The main environmental factors affecting seed production are temperature, light, humidity, rainfall and irrigation, soil, wind and location.

(a) **Temperature:** It is a vital factor affecting vegetable seed production influencing germination, flowering, pollination, seed setting, seed ripening, seed yield, seed quality and seed storage. Germination and growth are affected by prevailing temperature in soil during sowing time. Vegetable crops are grouped into three categories on the basis of their temperature requirement for growth and development:

i. Low temperature range (7-13^{0}C): Cole crops, root crops, lettuce, spinach, etc.

ii. Moderate temperature range (13-18^{0}C): Tomato, brinjal, sweet pepper, etc.

iii. High temperature range (18-24^{0}C): Okra, cucurbits, etc.

Some vegetables require low temperature stimulus before they can flower (*the process is called vernalization*). Many biennial vegetables like cabbage, beet, Brussels sprouts, European types of carrots have vernalization requirement.

Viability and receptivity of stigma are reduced due to excessive high temperature. Poor fertilization and abortion of flower may occur at high temperature.

(b) **Light:** It has direct effect on photosynthesis and daylength. Light comes from sun and provides energy for formation of sugar and related compounds. According to photoperiodic requirement, the crop plants are grouped into three, namely, *short day plants* which require 8-10 hours dark period for induction of flowering (*e.g.*, radish, spinach, beet, etc.), *long day plants* in which the flowering is induced by 10-14 hours dark period (*e.g.*, sweet potato), *day neutral plants* in which the flowering is induced irrespective of duration of light (*e.g.*, tomato, pumpkin, broad bean,etc.).

(c) **Humidity:** Relative humidity affects plant growth by influencing transpiration rate. In some vegetables (*e.g.*, beans) low humidity seems to increase abortion of flowers, decrease viability of pollen in some plants. High atmospheric humidity is one of the reasons for incidence of some fungal diseases in many vegetable crops. Activities of pollinating insects increase under optimum humidity.

(d) Rainfall and irrigation: The weather should be dry and warm under rainfed condition during finishing the crop. Many plants have moisture sensitive stages during germination, flowering and fruit setting. Sufficient soil moisture should be available during these stages. Irrigation depends on soil types, nature of crops and weather condition. Excessive rainfall is responsible for higher incidence of diseases and makes harvesting operation difficult. Reduction in viability of harvested seed and premature germination are happened due to high rainfall during seed ripening period.

(e) Soil: Soil should be well levelled, free from weeds, soil borne diseases and pests. The same crop should not have been grown in the preceding season. The soil should have good mechanical support, water and nutritive capacities, drainage facilities and ability to allow gaseous exchange for root respiration. The soil should be fertile as well as productive and free from acidity, alkalinity and water logging.

(f) Wind: High wind increases water loss from soil and crops due to evaporation and transpiration, carries pollens over a long distance, interferes with the activities of pollinating insects, increase loss of seed by shattering during ripening. So the areas inclined to high wind intensity should not be selected for seed production.

(g) Location: Regions with ample sunshine, moderate rainfall, free from strong winds and absence of extremes of hot and cold conditions should be selected for quality seed production of vegetable crops.

2.4 Field standards

These standards are carried out for maintenance of genetic purity. Important safeguards in this regard are:

i. **Preceding crop requirements:** It means that for the prescribed minimum period the same crop/any contaminating crop must not have been grown on the same field.

ii. **Isolation:** One major factor during seed production is to ensure that the possibility of cross-pollination between different cross-compatible plots or fields is minimized. In addition to this, adequate isolation also assists in avoiding admixture during harvesting and the transmission of pests and pathogens from alternative host crops. Seed crops can be isolated by time and by distance.

- *Isolation by time:* This type of isolation is possible within individual farms of multiplication stations. In this system, seed production is arranged in such a manner that the cross-compatible varieties are grown in successive years or seasons provided the rules regarding rotation are applied.

- *Isolation by distance:* When isolation by time is not possible, then isolation by distance is to be followed. Isolation distance primarily depends on the nature of pollination of the crop.

A minimum isolation distance is provided to prevent out crossing between different cultivars of the same crop and cross compatible species to prevent genetic contamination and seed borne disease contamination from the neighbouring fields. Isolation distances for foundation and certified seed production of vegetable crops are maintained as per seed certification standards.

Cross compatible species required to be isolated.

Sl No	Vegetable crops	Minimum isolation distance(m)		To be isolated from
		Foundation Seed	Certified Seed	
1.	Amaranthus	400	200	Same variety, wild spp
2.	Tomato	50	25	Same variety, other spp
3.	Brinjal	200	100	Same variety, other spp
4.	Capsicum/Chilli	400	200	Same variety, other spp
5.	Cauliflower, cabbage, knolkhol, Brussels sprout, sprouting broccoli, kale	1600	1000	Other cole vegetables and varieties
6.	Chinese cabbage	1600	1000	Turnip
7.	Amaranthus	400	200	Wild amaranthus
8.	Lettuce	50	25	Wild lettuce
9.	Okra	400	200	Wild species
10.	Beet	1600	1000	Garden beet, swiss chard and sugar beet
11.	Cucurbits	800	400	Wild cucurbits and other varieties, musk melon from long melon and vice versa, pumpkin from winter and summer squashes and vice versa
12	Radish, Turnip	1600	1000	Same variety, Chinese cabbage, rape, mustard, and rutabaga for turnip
13	Fenugreek/Methi	50	25	Other varieties
14	Onion	1000	400	Other varieties

iii. Field inspection: Principal objective of field inspection is to conform that quality seeds are produced which are of good seeding value and true to the type. Field inspection should be carried out by a well skilled and experienced person who have authorized by the certification agency. Several rougings are done at different times of the day and by walking in different directions. All off type, undesired and other crop plants, objectionable weeds and diseased plants should be removed from the seed production field. During seed production, generally, maximum four and minimum two field inspections are standardized for certification. If needed, verification of seed quality may be done after harvesting and seed testing also may be conducted. Genetic purity of seed is conformed if it is approved by the seed certification agencies.

2.5 Seed standards

Seed standards consist of the following:

1. Generally foundation or certified seeds are supplied to the seed growers. The seeds for raising a seed crop should be from approved source and genetically pure.

2. Pure seeds may be mixed with the seeds of other crop plants, volunteer plants, seeds of other varieties, weed seeds and inert matter due to mechanical admixture in the field and threshing floor. Minimum percentage of pure seeds and maximum permissible limits for seeds of other crops and varieties, weed seeds and inert matter have been prescribed.

3. Maximum permissible limits for seeds infected by seed borne diseases have been fixed to ensure good seed health.

4. Pure seeds have the ability to complete the process of germination and must have minimum germination percentage.

5. Seeds should be properly dried before storage to maintain optimum moisture level in it. Maximum permissible limits for moisture content in seed have been prescribed for safe storage of seed.

2.6 Seed harvesting

Vegetable seed crops are harvested at proper stage of maturity at 18-20% moisture content in most cases. Pre-mature harvesting results shriveling of seeds on drying. Reduction in seed yield due to shattering and less viability of seed are happened due to delay in harvesting.

Vegetable seed crops are broadly grouped into three on the basis of state of seeds and harvesting time.

1. **Dry seed:** Before harvesting seeds are dried on plant, *e.g.*, *Brassica*, peas, beans, onion, beet, spinach, carrot, lettuce, fenugreek, etc.

2. **Fleshy fruits:** Ripe fruits are picked from the plants and dried before seed extraction. Seeds are separated by opening the dried fruits, *e.g.*, okra, bottle gourd, sponge gourd, ridge gourd, tropical pumpkin, chilli, etc.

3. **Wet fleshy fruits:** Fruits of some seed crops contain high level of moisture. Some seeds have gelatinous or mucilaginous coating and can be removed by fermentation or acid or alkali treatment, *e.g.*, tomato, brinjal, cucumber, bitter gourd, etc.

Methods of harvesting

1. **Manually:** Fruits of solanaceous and cucurbitaceous vegetables, okra, chilli, seed heads of onion, carrot, inflorescence of *Amaranthus* are picked or cut with a knife or secateurs and kept in a bamboo basket or any other suitable container.

    Whole plants of some vegetable crops like *Brassica*, lettuce, radish are cut by a sickle or secateurs. Seed crops of peas and beans are harvested by putting the whole plants.

2. **Mechanical harvesting:** This system is followed for large scale seed production. A combine harvester can perform both cutting and threshing operations.

2.7 Seed threshing

This operation is done after thorough drying of the harvested materials on the smooth threshing floor or tarpaulin. Threshing operation is performed by hand, animals or machine.

i. **Hand threshing:** It may be done in the following ways:

    (a) **Rubbing:** Seeds are extracted by rubbing the seed heads with hand, *e.g.*, legumes

    (b) **Beating:** Seed materials are beaten with a wooden stick, *e.g.*, cole crops, legumes, leafy vegetables

    (c) **Flailing:** Seeds of dried fruits are separated by beating with a flail

    (d) Walked on: A person is directed to walk on the seed materials till the seeds are separated.

  ii. Threshing with animals or tractors: Harvested materials are dried in sun. Animals like cattle are assigned to trample the thick layer (10-15 cm) of plant materials in a circular fashion till the seeds are separated. Threshing with tractor is carried out in a same manner. This method is followed in the areas producing seed in a large scale.

  iii. Machine threshing: Threshing machine has a cylinder revolving in a concave. For threshing of vegetables/spices seeds, speed of cylinder should be 1100 revolutions per minute (rpm). Speed of 700 rpm is used for threshing large seed legumes.

Winnowing

After extraction, seeds are cleaned by blowing against wind or fan to remove chaff, trash, light materials etc. Winnowing machine can be used for the purpose. Sometimes sieving is required to remove stone, soil, etc.

2.8 Seed drying

Seed drying is essential as it reduces moisture content of seed to safe limits and maintain viability, vigour and protect from attack of mould, bacteria and pests. Beside seed drying concedes early harvesting, long-term storage of seeds, efficient use of land and manpower, use of plant stalk as fodder and production of quality seeds.

Methods of seed drying are grouped into two; i) sun drying ii) forced air drying. In sun drying, seeds are dried by spreading the seed in a thin layer (2-3 cm) on floor or tarpaulin under solar radiation. Frequent stirring of seeds is essential for rapid and uniform drying. In this process seed germinability can be affected due to high temperature and ultraviolet radiation when moisture content is high. Forced air drying method is required for large scale seed production. In this system natural air or heated air is blown through a layer of seed until drying is completed. Two types of drier, *viz.*, continuous flow drier and batch drier are used in this system.

Some vegetable seeds dry quicker than others. According to drying rate vegetable seeds are divided into quick driers (lettuce, cucurbits), medium driers (tomato, carrot, beet) and slow driers (*Brassica*, peas, beans, onion).

For open storage. moisture content will be 12% for starchy seeds and 9% for oily seeds. Moisture percentage in seed should be 6-8 when stored in sealed containers.

2.9 Seed treatment:

It is the application of fungicides, insecticides or their combination (if compatible) to seed with the view to disinfect and disinfest them from soil or seed borne pathogens and storage insects. Seed treatment also includes subjection of seeds to solar energy exposure and immersion in conditioned water.

Seed disinfection refers to eradication of fungal spores within the seed coat or in inner tissues. Fungicidal treatment will be effective for the purpose. Seed disinfestations is the destruction of surface organisms (fungi, bacteria, insects) that have contaminated but not infected the seed surface. Chemical dips, soaks, fungicides applied as dust, slurry or liquids provide satisfactory control. The purpose of seed protection is to protect the seeds and young seedlings against diseases and pests present in soil or air borne organisms when seedling emerge.

2.10 Seed certification

The objective of seed certification is to ensure genuineness and quality of seeds to the seed growers. The purpose of this programme is to maintain and make available to the growers high quality seed of genetic identity and purity.

Seed certification programme are conducted by the statutory bodies and various agencies under Seed Act 1966 and are under (i) Central Seed Committee, (ii) Central Seed Certification Board, (iii) State Seed Certification Agencies, (iv) Central Seed Testing Laboratory, (v) State Seed Testing Laboratory.

Seed certification procedure

(a) **Verification of seed source:** The seed grower must submit documentary evidence to the certification agency regarding the source of seed for production of foundation or certified seed.

(b) **Field inspection to verify conformity to the prescribed field standards:** Field inspection programme are performed to verify identity of variety, cropping history of seed plot, land requirements, isolation distance and to check weeds, off types and seed borne diseases. Field inspections are also necessary during sowing, vegetative and flowering stages.

(c) Supervision at pre- and post-harvesting stages: Seed crop should be inspected during harvesting, threshing and cleaning. Besides it is necessary to check threshing floor, seed containers and sampling arrangements. After threshing, cleaning and drying seeds are treated with fungicides and insecticides. Seed lots are then bagged with proper identification marks.

(d) Seed sampling and testing: Representative samples of seed are subjected to analysis in central or state seed testing laboratories for germination, purity, moisture, other crop and weed seeds and seed health to ensure whether the seed lots meet the requisite seed certification standards.

(e) Tagging and sealing: On the basis of report of field inspection and seed testing in laboratory the certification agency tags. Normally white tags are issued for foundation and blue tags for certified seeds. The certification tag is affixed on bag or containers and sealed in a proper way.

2.11. Seed vigour

According to the Association of Official Seed Analysis (AOSA), seed vigour comprises those seed properties which determine the potential for rapid, uniform emergence and development of normal seedlings under a wide range of field conditions. Vigour tests of a seed lot prior to sale are equally important to purity and standard seed germination test. In addition to seed germination test, vigour test of seed, for getting accurate information concerning a seed lot's field performance, is essential for the following reasons:

i. Seed germination essentially emphasizes on seedling morphology leading to the production of normal plant and not on rapidity of growth as a primary potential for successful stand establishment

ii. The germination percentage is the sum of strong and weak seedlings as recorded by an initial count (strong seedlings) and a final count after a certain interval (weak seedlings). But the weak seedlings rarely show a successful field emergence.

iii. A seed is considered either germinable or not *i.e.*, germination is scaleless. Whereas, seed vigour documents the quality of the seed leading to its stand establishment.

iv. Compared to vigour test, germination tests must be conducted on artificial, standardized, essentially sterile media in humidified, temperature controlled chambers. These conditions are so synthetic that they seldom relate to field situations.

2.11.1 Seed vigour test

The main challenge of vigour test is to identify one or more quantifiable parameters that are common to seed deterioration. Most vigour tests have focused on measuring the biochemical and physiological changes known to occur during seed deterioration. The characteristics of seed vigour test, as described by Macdonald (1980) are as follows:

i. The test should be *inexpensive* involving minimum investment of labour, equipment and money.

ii. The test should be *rapid* to provide the seed producers a quick information on seed quality for availing a competitive marketing advantage.

iii. The test should be *uncomplicated* and simple so that they can be easily performed in the laboratories without involving additional labour and specialized skill.

iv. The vigour test should be *objective*, avoiding sorts of subjective interpretation to a considerable extent.

v. The outcome of the test must be *reproducible*.

vi. Vigour test should essentially be *correlated with field performance*.

The most commonly conducted vigour tests include cold test, accelerated ageing (AA) test, electrolytic leakage test and seedling growth rate test etc.

(a) Cold test: It is one of the oldest methods of seed vigour test and the most preferred one. Seeds are placed in boxes, trays or rolled towels that contain field soil (having 60-70% water holding capacity)and held at 10^0C for 7 days before being moved to 25^0C (ideal condition). The numbers of seeds that emerge are counted after 4 days. The ability of seeds to germinate and emerge in cold wet soil is affected by genotype, physical and physiological quality of seed, microorganisms (seed/soil) and seed treatment. The combined effects of all these factors are measured by the cold test. Usually it represents the lowest emergence that would be expected from a seed lot when planted under reasonably satisfactory field conditions, while the germination test represented the highest emergence potential that could be expected.

The general principles of the cold test have been applied successfully to corn, sweet corn, soybean, sorghum and other kind of seeds. One of the major difficulty faced by the test is the lack of uniformity in field soil, mainly differing in moisture, pH, texture, presence of microorganisms etc. which produced varying results.

(b) Accelerated ageing (AA) test: It is a common test for the agronomic and vegetable seeds. In this case, before standard seed germination test, seeds are subjected to high temperature (40 to 45^{0}C) and high relative humidity (around 100%) for short periods (3-4 days). After then the seeds are removed from the stress conditions and placed under optimum germination conditions.

This test exposes seeds for short periods to the two environmental variables that cause rapid seed deterioration; high temperature and high humidity. High vigour seed lots will withstand these extreme stress conditions and deteriorate at a slower rate than low vigour seed lots.

(c) Electrolytic leakage test: Electrolytic leakage is a phenomenon which occurs during seed deterioration. Usually, the electrical conductivity can be measured by using a conductivity meter. In case of large seeded crops, however, the measurements are corrected with field emergence.

(d) Seedling growth rate test: After finding out the percentage germination with the help of a standard germination test, this method is started. In this case, determination of shoot and root length or seedling weight is done after a certain period of time at controlled temperature. In this way strong and week seedlings from a seed lot can be sorted out. Of late, Ball vigour index, which employs computer analysis of video images of all seedlings in plug trays after a certain time, was introduced by Ball Seed Company (West Chicago, IL) is suggestive of seedling greenhouse performance.

The seed vigour tests must be conducted on a number of seed lots simultaneously with the inclusion of a control seed lot with known seed vigour to serve as reference because the test result do not predict percentage field emergence due to varying environmental conditions differing from field to field, day to day and year to year. It is also difficult to correlate high seed vigour with higher yields because of insufficient evidence to prove it. Future research orientation may be planned in this direction.

2.12 Some important facts related to seed

2.12.1 Seed purchase

During purchase of seed the following points must be noted:

- Seeds should be reliable for freshness and purity.
- Seeds must be packaged for the current year containing information on expected percentage of germination.
- The ideal germination% of most seeds is 65-80%.
- 65-75% of the germinated seeds grow into satisfactory seedlings.
- The selected seeds must meet the requirement for size, colour and growth habit.
- Apart from germination%, the seed packet should inform on country of origin, bloom time, specific germination requirements, cultural requirements and disease resistance.
- Some seed packets indicate whether seeds are chemically treated or not.

2.12.2 Seed collection

- Often the home gardeners collect and save seeds from the randomly pollinated plants which may not be identical to the parent plants.
- Sometimes they save seeds of hybrid cultivators for next year use.
- Self-pollinated non-hybrid pure bred annual vegetable seeds that can be saved excellently include lettuce, beans, peas, herbs etc.
- Collection of few seeds which are not available commercially may help the gardener to maintain varieties.
- Collection of seeds or planting materials from foreign countries may invite certain serious insects and diseases.

2.12.3 Seed storage

- Once seeds are dried completely, they should be placed in airtight containers marked with name and date of collection.
- Usually seeds are stored at 45^0C with low humidity (as in a refrigerator).
- Properly stored seeds may remain viable upto 5 years.
- Some seeds should be checked to study the germination rate before sowing/planting.

2.12.4 Seed germination

- Moisture, light (or darkness), oxygen and heat play a part in triggering germination.

- Even with ideal conditions, some seeds are still very difficult to germinate.

- Water penetrates the seed coat, causes the endosperm to swell, dissolves nutrients in the endosperm making them available to the embryo, growth begins-all expressed as germination.

- Light can stimulate of inhibit germination.

- Seeds must respire to breakdown the food stored in seed, thereby necessitating oxygen.

- Every seed has an optimum temperature range for germination.

2.12.5 Seed scarification

- Seed coat of certain seeds being extremely tough, must be penetrated by special means *e.g.,* breaking, scratching or softening to allow moisture. This is known as scarification.

- Scarification may be of two types-mechanical (with sand paper or hammer) and hot water (in 75-100°C hot water for 12-24 hrs).

2.12.6 Seed stratification

- Some seeds need to be exposed to a period of low temperature (chilling) and moist conditions for breaking their dormancy. This is stratification.

- Many trees, shrubs and certain perennials typically require stratification.

- Stratification can be accomplished by placing a seed in a moist, sterile (pasteurized) growing medium, such as a mix of equal parts clean sand and peat or sphagnum peat moss in a disinfected container. The container is enclosed in a tightly sealed plastic bag and placed in the non-freezer section of the refrigerator until sprouting.

- Some seeds may require scarification alongwith stratification to germinate.

2.12.7 Seed sowing

- Seed sowing may be of two types *viz.*, sowing seed indoors and sowing directly into the garden.

- Sowing indoor is an easy and cheap method of producing vegetables, annual flowers and some perennial plants. But it demands fluorescent or grow lights, disinfested containers with excellent drainage, pasteurized (sterile) seed starting medium, proper temperature, adequate ventilation and constant care in transplanting, watering, lighting, fertilization, pinching, if any and hardening of seedlings before transplanting into main garden.

- On the other hand, many flowers and vegetables may be sown directly into the garden. It avoids transplant shock, requires less work, but involves more risk from weather, pests, diseases and erosion.

2.12.8 Seed Quality Control

- The minimum limits of germination and purity standards have been notified for different kinds [Section 6 (a) of the Seeds Act 1966]

- A mark or label should be attached with the seed container with required particulars [Section 6 (b) of the Seeds Act 1966].

- The sale of seeds of notified kinds and varieties are regulated. Every person selling seeds should sell seeds

  ❖ Which are identifiable to its kind and variety.

  ❖ Should posses the minimum required seed standards for germination, physical purity and genetic purity.

  ❖ Should affix a mark or label with correct particular to the seed container.

  ❖ Should carry out such other instruction given by the State Government as prescribed under rules [Section 7 of the Seeds Act 1966].

Also under Rule -13 there are some requirements for every dealer to comply with

1. No. person shall keep any seed for sale after the date of its validity

2. The mark of label should not be tampered

3. Every person selling seeds should keep complete lot wise records of seeds for atleast for a period of three years which also include a seed sample from each lot which may be kept for one year.

Maintaining the quality of seeds – Farmers' Role

Farmers should purchase the seeds only in the licensed seed selling points. They must verify the details furnished in the producers label attached to the seed bag. The label should possess the following details

- SL.No.
- Crop
- Variety
- Lot No.
- Date of Test (Date/Month/Year)
- Validity period (Date/Month/Year)
- Germination%(Minimum)
- Physical purity (Minimum)
- Genetic purity (Minimum)
- Net weight
- Chemical used for seed treatment
- Name and address of the producer

   (Caution in red ink should be furnished for Name of the chemical, and "Do not use for Food, Fodder, or Oil purpose").

Certified seed Tag

The farmers must purchase seeds of verified variety and not the un-notified ones.The farmers should report the matter to the concerned seed inspector in case they suspect anything about the quality. Further they should keep with them the tag, label and the container of the seed bag. In case of germination failure, the fact should be reported to the concerned seed inspector. On each purchase, receipt must be ensured without fail bearing details on variety, lot number and validity period.

Seed testing

Seed testing is the science of evaluation and verification of the seeds with regard to their planting value. The primary aim of this is to obtain accruable and reproducible results regarding the quality status of the seed samples submitted to the Seed Testing Laboratories.

Service samples

These types of samples are drawn and sent to the seed testing lab by the Farmers, Producers and the Sellers to know their seed standards.

The samples received from the three categories are analysed and the outcomes are communicated to the concerned within 30 days from the date of receipt of the samples.

Certified samples

These are for certification purposes by the producers. This type of samples are drawn by the certification officers who send them to the seed testing lab through the Assistant Director of Seed Certification with a secret code number. On receipt of the results, the code will be decoded by the ADSC and then the results are sent to the producer for further certification procedure to be completed.

Official samples

These are for the quality control purpose. This type of samples are drawn by the seed inspectors for the quality control as per the seed control order and Seed act to assure the availability of quality seeds to the farming community.The seed samples received in the Seed Testing Laboratories are analysed and results are communicated to the concerned. The following tests are conducted:

- **Germination**
- **Physical purity**
- **O.D.V. (Other Distinguishable Variety)**
- **Moisture**

The Seed Testing Laboratory is the hub of seed quality control. Seed testing services are required from time to time to gain information regarding planting value of seed lots. Seed testing is possible for all those who produce, sell and use seeds (**Minimum seed standards**).

2.12.9 Seed Export / Import in India

The export/import of seeds and planting material is governed by the Export and Import (EXIM) Policy 2002-07 and amendment made therein. Restrictions on export of all cultivated varieties of seeds have been removed w.e.f. 01.04.2002, except the following:

(i) breeder or foundation or wild varieties; (ii) onion, berseem, cashew, nux vomica, rubber, pepper cuttings, sandalwood, saffron, neem, forestry species and wild ornamental plants; (iii) export of niger which is canalized through TRIFED, NAFED, etc. (iv) groundnut, exports of which is subject to compulsory registration of contract with APEDA;

The export of these seeds is restricted and is only allowed on case-to-case basis under licence issued by Director General Foreign Trade on the basis of the recommendations of Department of Agriculture and Cooperation.

The provisions regarding import of seeds and planting material are as under:

(a) Import of seeds/tubers/bulbs/cuttings/saplings of vegetables, flowers and fruits is allowed without a licence in accordance with import permit granted under Plant Quarantine (Order), 2003 and amendment made therein. (b) Import of seeds, planting materials and living plants by ICAR, etc. is allowed without a licence in accordance with conditions specified by the Ministry of Agriculture; (c) Import of seeds/tubers of potato, garlic, fennel, coriander, cumin, etc. is allowed in accordance with import permit granted under PQ Order, 2003. (d) Import of seeds of wheat, rye, barley, oat, maize, rice, millet, jowar, bajra, ragi, other cereals, soybean, groundnut, linseed, palmnut, cotton, castor, sesamum, mustard, safflower, clover, jojoba, etc. is allowed without licence subject to the New Policy on Seed Development, 1988 and in accordance with import permit granted under PQ Order, 2003.

The EXIM Policy reiterates that all imports of seeds and planting material would be regulated under the Plant Quarantine Order 2003.

As per World Seed Trade Statistics, India has sixth largest size of domestic seed market in the world, estimated to be at about 1300 million dollars. However, India's share in global trade in seeds (import and export) is of only about 37 million dollars only. To give a boost to seed export, India has decided to participate in OECD Seed Schemes for the following categories of crops (http://seednet.gov.in):

- Grasses and legumes
- Crucifers and other oil or fibre species
- Cereals
- Maize and sorghum
- Vegetables

Chapter 3
Seed Production (Vegetable Crops)

3.1 Tomato

Botanical name   :   *Lycopersicon esculentum* Miller

Family   :   Solanaceae

Chromosome No.   :   2n=24

Tomato is an important and popular vegetable cultivated all throughout the world. It is normally grown as an annual crop both for seed and vegetable as well. But in tropical South America, the native place of the crop, tomato is grown as perennial. It can also be grown in greenhouse or as soilless culture or hydroponics. It is one of the most important protective food crops of India. It is grown in 0.458 M ha area with 7.277 M mt production and 15.9 mt/ha productivity. The major tomato producing states are Bihar, Karnataka, Uttar Pradesh, Orissa, Andhra Pradesh, Maharashtra, Madhya Pradesh and West Bengal. In West Bengal, tomato is grown over an area of 43,600 ha with the production 0.588 M mt and productivity of 13.6 mt/ha. Tomato is rich source of vitamins A, C, potassium, minerals and fibers. Tomato is consumed fresh as salad, as an important ingredient for the preparation of popular Indian cooked dishes or processed in the form of paste, canned or dehydrated product (NCPAH,2013).

3.1.1 Botany

The plant is branched with alternate pinnate compound leaves.

Initially the stems are round, soft, brittle and hairy which gradually become angular, hard and semi-woody. Cluster of flowers (inflorescence) borne on lateral axes. When flower clusters are present in each third internodes of a main axis and plant growth at the apical end continues with favourable environment, it is known as 'indeterminate' type. Whereas, the type in which flower clusters occur more frequently, sometimes at every node, until the formation of a terminal cluster, is called as 'determinate' or 'self-pruning' type. The later types have a quicker tendency of maturity than the former.

In most varieties the number of flowers per cluster is 4 to 5 or even more. Flowers with 5-10 sepals and 5 or more yellow petals united in a short tube. The stamens, 5 in number, are attached to the base of the corolla tube. There is no definite flowering peak and anthesis appears to be correlated with temperature and soil moisture. Stigma is receptive at the time of anthesis but anthers do not dehisce until about 24-28 hours later. Tomato is essentially a self-pollinated crop but cross pollination up to the extent of about 5% may occur through insects *viz.*, bees. Fruit is a berry with fleshy placenta. Seeds are flattened, pubescent, the number of which vary from 150 to 300 or even more per fruit. The test weight is approximately 2.4g.

3.1.2 Environmental and isolation requirements

Tomato is very much sensitive to frost and does not stand prolonged periods below 10^0C and above 25^0C. Moreover, high relative humidity and high temperature for a prolonged period may invite bacterial canker (*Corynebacterium michiganense*), a common major pathological problem of tomato. A soil with adequate liming material having pH value of 6.5 is optimum. Light textured soil is preferred for earliness. A manurial dose of NPK @ 1:2:1 ratio is to be applied during final land preparation.

There must be a gap of minimum two years between a tomato seed crop and any previous tomato or related crop; otherwise the soil is to be partially sterilized. Longer rotations are necessary to avoid the attack of disease (*e.g.*, wilt) and insect (*e.g.*, nematode). For foundation and certified seed production the required minimum isolation distance between crops are 50 m and 25 m, respectively (Agarwal, 1980).

3.1.3 Cultural requirements

The cultural requirement of tomato for seed is more or less identical with that for open market tomato or tomato for processing. Usually seedlings are raised on seedbeds and transplanted to the main seed production field only after the emergence of approximately three leaves. Sometimes protective structures like cold frames or hot beds are preferred to grow as well as to protect the young seedlings from frost and also prepared in the areas with a relatively short growing season.

Land is prepared with smooth surface free from clods by thorough ploughing, harrowing and levelling. For transplanting in one hectare of land, seedlings are produced by sowing 60 g of seed approximately in 250 m^2 of seed bed. Seeds are to be treated with *Trichoderma viride* 4 g or *Pseudomonas fluorescens* 10 g or Carbendazim 2 g per kg of seeds 24 hours before sowing. Just before sowing, the seeds are treated with *Azospirillum* @ 40 g / 400 g of seeds. Then sowing is done in lines at 10 cm apart in raised nursery beds and cover with sand. Immediately after sowing, the beds are irrigated. Protection of seedlings from direct sun, heavy rains and frost is extremely essential. The nursery area is generally covered with 50% shade net and the sides are covered using 40/50 mesh insect proof nylon net. Transplanting is usually done within 3-4 weeks after sowing. In the plains, June to November is the normal growing season. Indeterminate type of plants are planted approximately 50 cm apart, with 1 m between the rows. The determinate (bush) types are planted 45-120 cm within the rows and 90-150 cm between the rows depending on varieties.

The young seedlings need regular watering but water logging at any stage of crop growth is harmful for the crop. The irrigation interval during summer months should not exceed a week, whereas in the winter it may be 2-3 weeks. Heavy irrigation after a long dry spell must be avoided as it often causes fruit cracking. Salunkhe *et al.,* (1987) recommended the following:

Crop	FYM (t/ha)	Amm Sulfate (kg/ha)	SSP (kg/ha)	MOP (kg/ha)
Early(rainy) Season crop	5	500	312	85
Winter crop	5	400	250	68
Spring-season crop	5	300	188	50

To control weeds, regular hoeing is essential in the field. Application of Pendimethalin 1.0 kg a.i./ha or Fluchloralin 1.0 kg a.i/ha as pre-emergence herbicide is advocated, followed by hand weeding once at 30 days after planting. The indeterminate cultivars require staking and/or tying. Off-type plants should be removed to avoid cross pollination. At maturity, plants with fruits of unwanted external and internal characteristics (*e.g.*, shape, colour, size, nutritional value etc.,) should be removed. Care must be taken to control the prevailing pathogens and pest. In India, the major diseases of tomato are damping off (*Pythium* sp., *Phytophthora* sp. and *Rhizoctonia* sp.), early blight (*Alternaria solani*),fruit rot (*Phytophthora* sp.), Fusarium wilt (*Fusarium lycopersicae*),Powdery mildew(*Leveillula taurica*), and viral diseases (*viz*, tomato mosaic and leaf curl). Good drainage in the nursery bed, drenching seed beds with Bordeaux mixture and spraying seedlings with Bordeaux mixture or other copper fungicides often keeps away the pathogen. Among the important pests of tomato, the important ones are aphid, white fly, epilachna beetle, fruit borer, nematodes, etc. For controlling fruit borer Azadirachtin 1.0% EC (10000 ppm) @ 2.0 ml/l or Quinalphos 25% EC (1.0 ml/l) and for managing whitefly spraying of dimethoate Dimethoate 30% EC (1.0 ml/l) or Malathion 50% EC (1.5 ml/l) or Oxydemeton –Methyl 25% EC (1.0 ml/l) are found effective. Soil application of *Bacillus subtilis* (BbV 57) or *Pseudomonas fluorescens* as seed treatment @ 10 g/kg of seeds and soil application @ 2.5 kg / ha for the management root knot and reniform nematode infestation in soil and root. Application of liquid formulation of *Bacillus subtilis* (BbV 57) or *Pseudomonas fluorescens* @ 1000 ml/ha through drip irrigation for the management of root knot nematode in tomato is also recommended.

3.1.4 Harvest and Post harvest management

Ripe fruits are collected by hand over a period of time whenever it ripes or by a single once-over operation using a specially designed mechanized harvester. The indeterminate types are often harvested by hand where adequate labour is available or where the cultivars are with sequential ripening. Care should be taken to avoid fruit damage during harvesting. The fruit while grasped in hand, should be dislodged from the vine by twisting, with the thumb kept pressed against the vine.

Post harvest operations include extraction, washing, drying, further cleaning and packing of seeds. Seeds are extracted with seed extractor by separating the gelatinous coated seed from the crushed fruit debris. Usually the crushed material is passed into a revolving cylindrical screen, which allows the seeds and juice to pass through the mesh, while the fruit debris passes through the cylindrical screen to drop in the field. The next step is the removal of gelatinous layer from the extracted seeds either by fermentation or by extraction with acid or sodium carbonate. In fermentation method, the seed and pulp mixture are allowed to ferment without addition of water for upto 4-5 days in non-metallic containers (to avoid staining of seeds). The process disintegrates the mucilaginous matter adhering to the seed, allowing the seed to sink to the bottom. Covering of the containers (check the fruit flies) and periodical stirring of the material (to ensure uniformity in the container as well as to prevent fungal attack at the surface thereby avoiding discolouration of the seeds) is necessary. Long-term fermentation (at 25^0C) may decline the germination percentage severely. Extraction with acid involves adding hydrochloric acid to pulp @ 35 g/11.34 kg of material and mixing thoroughly. The process takes only about 15-30 minutes. Sodium carbonate (10%) may also be used as an alternative to hydrochloric acid. The volume of seed-pulp mixture and the volume of sodium carbonate is to be equal. If the mixture is kept at ambient temperature for upto 2 days, a satisfactory separation of seed and pulp can be expected.

After extraction of seeds, they are washed thoroughly with a shaker-washer (It consists of two screen with a tray in each one suspended above the other in the same frame. Pulp and seed are periodically placed on the upper screen which prevents the passage of the coarse pulp and skin, allowing the seed and smaller particles. The lower screen is removed and seeds are allowed to drain only after sufficient seed accumulated on it.) or by involving the use of current box, washing flume, or sluiceway (particularly when large quantities of seed need to be washed).

A common method of drying is spreading of seed in screen-bottom trays placed on racks outdoor so that air passes both above and below the screens. Stirring of the seed at certain interval increases the drying speed. Artificial drying is common in areas of high humidity. Generally the cabinet and revolving screen driers with forced warm air are used.

Normally tomato seeds do not require further cleaning where the extraction process has been done efficiently. However, it can be done with an air-cleaner or screen-cleaner. When different groups of seed adhere together, the seed lot is brushed or rubbed to separate the individual seeds.

To get the highest viability and germinability upto about 8 years, storing of tomato seeds in glass jars is essential. The next best packages are polythene boxes and plastic bags. In commercial field production of tomato, the rule is that seed weight should be 1% of the fresh fruit weight; e.g., one ton fruit yield should produce 10 kg of seeds. Thus seed yield will depend on yield of fresh fruits. In India, Singh *et al.*, (1964) recorded an average yield of 145 kg/ha of tomato seed.

3.2 Brinjal

Botanical name : *Solanum melongena*
Family : Solanaceae
Chromosome No. : 2n=24

Brinjal or egg plant is a widely grown crop in tropics, subtropics and warmer temperate areas of the world. It is thought to be originated in India and china. The crop is extensively grown in India, Bangladesh, Pakistan, China and Philippines. Though the international trade is negligible, brinjal holds a premier position in domestic market and is consumed as a cooked vegetable in various way. India produces about 7.676 M mt of brinjal from an area of 0.472 M ha with an average productivity of 16.3 mt/ha. The brinjal producing states are Orissa, Bihar, Karnataka, West Bengal, Andhra Pradesh, Maharashtra and Uttar Pradesh. In West Bengal brinjal is grown in 0.140 M ha area with the production of 2.388 M mt and productivity of 17.0 mt/ha. The major brinjal producing belts in West Bengal are Hoogly, 24-Paraganas and Burdwan. It is quite high in nutritive value and possesses some medicinal properties also (e.g. used against diabetes, toothache, liver complaints etc. In the past, farmers maintained and supplied seeds of brinjal with special type of varieties adapted in the region. Brinjal has ayurvedic medicinal properties and white brinjal is good for diabetic patients. Now there are an increased number of F_1 hybrid varieties bred by seed companies and the seed production of brinjal is shifting from farmer's hands to seed companies (NCPAH, 2013).

3.2.1 Botany

Brinjal is a bushy plant with large, alternate leaves and large violate coloured, perfect, solitary or clustered flowers. Calyx five-lobed, corolla expanded and purple, anthers arranged around the style beyond which the stigma can be seen. Depending on the length of styles, there are four types of flowers, *viz.*, (i) long styled with big size ovary (25%,

(ii) medium styled with medium size ovary (10%), (iii) pseudo-short styled with rudimentary ovary (15%) and (iv) true short styled with very rudimentary ovary (50%). Fruit production is evident in long and medium styled flowers, whereas, pseudo-short and true short styled flowers do not set any fruit. Chances of cross pollination are more in long-styled flowers. The varieties of *Solanum melongena* L. display a wide range of fruit shapes and colours, ranging from oval or egg-shaped to long club-shaped; and from white, yellow, green through degrees of purple pigmentation to almost black. There are three main botanical varieties under the species *melongena*. The round or egg-shaped cultivars are grouped under var. *esculentum,* the long slender types are included under var. *serpentintum* and the dwarf brinjal plants are put under var. *depressum.*

3.2.2 Environmental and isolation requirements

Brinjal is sensitive to frost and requires a long warm growing season. Cool night and short summer is not favourable. Day temperature ranging 25-35 ^{0}C and night temperature of 20-27 ^{0}C is found to give satisfactory yield of seed. There is no specific daylength requirement. In most eggplant-growing areas, anthesis and pollen dehiscence in brinjal flowers occur between 6:00 and 11:00 in the morning. Pollen variability is retained for 8–10 days at a temperature of 20–22 °C and with a relative humidity of 50–55%. Best results are attained when pollen is used within four days.

Brinjal is a normally highly self-pollinated crop. The cone-like formation of anthers favours self pollination; but since the stigma ultimately projects beyond the anthers, there is an ample opportunity for cross-pollination. The rates of natural cross pollination may vary depending on genotype, location, and insect activity. It has been reported that the extent of natural outcrossing was from 2 to 48% in brinjal varieties in India and from 3 to 7% in China. At AVRDC, from 0 to 8.2% of natural outcrossing rates (with a mean of 2.7%) have been observed.

Neutral soil with a pH of 6.5 is the best. Acidic soil is to be ameliorated with adequate liming materials. The crop responds well to a basal fertilizer application with an NPK ratio of 1:1:1.

All the crops which are near relatives to brinjal should never be rotated with brinjal to avoid many soil-borne pests and pathogens. A period of four years is safe between two brinjal crops. Appropriate isolation of the seed-producing field from other varieties of the same crop is generally required. Being essentially self-pollinated, an appreciable amount of cross-pollination (through insects) also occurs in brinjal. For this reason, an isolation distance of at least 200 m for foundation seed and 100 m for certified seed is recommended.

3.2.3 Cultural requirements

The cultural requirements of brinjal for seed production are more or less the same as for fresh market vegetables all throughout the cultivation period excepting only in the seed bed. Seedlings are generally raised in seed beds with 500-750 g seed to transplant one hectare of land. Area required for raising seedling for planting 1.0 ha is 100 sq.m. Application of FYM 10 kg, neem cake 1 kg, VAM 50 g, enriched super phosphate 100 g and furadon 10 g per square metre before sowing is necessary. The seeds are treated with *Trichoderma viride* @ 4 g / kg or *Pseudomonas fluorescens* @ 10 g/kg of seed. The seeds are also treated with *Azospirillum* @ 40 g/400 g of seeds using rice gruel as adhesive. Irrigation is done with rose can. In raised nursery beds, the seeds are sown in lines at 10 cm apart and cover with sand. The nursery area is generally covered with 50% shade net and the sides are covered using 40/50 mesh insect proof nylon net. Transplanting of the seedlings is done at 30 – 35 days after sowing at 60 cm apart in the ridges. The young plants, carefully lifted avoiding any root injury, are transplanted when approximately six weeks old with 45 cm × 60 cm to 60 cm × 75 cm for long and round varieties and 75 cm X 120 cm for high yielding varieties. However, the plant density depends on vigour of the cultivar and the system of irrigation. The crops are kept free of weeds particularly in the early stages of crop establishment. The Indian Council of Agricultural Research (ICAR), New Delhi, has recommended 40-60 kg N, 60-80 kg P_2O_5 and 100-120 kg K_2O per hectare. The irrigation interval should be 2-3 days during summer and 12-15 days during the winter. Roguing for the seed crop is usually done at three stages. The first roguing is done in the early stages before blossoming to remove all the off-types concentrating mainly on general plant habit and foliage characters.The second is done at the start of anthesis after checking all the characters considered under first roguing adding only the spine characters. The third and final roguing, however, is done when the first fruits are fully developed. During this stage fruit characters including yield potential are brought into consideration.

Care must be taken to prevent the crop from pests and diseases. The most common dreaded pests of brinjal are fruit and shoot borer

(*Leucinodes orbonalis*), stem borer (*Euzophhera perticella*), jassid (*Empoasca* sp), and Epilachna beetle etc. For controlling fruit and shoot borer, a spray of Chlorpyrifos 20% EC (1.0 ml/l) or Dimethoate 30% EC (7.0 ml/10 l) at 10-15 days interval is found effective and a spray of Phosphamidon 40% SL (1.5 ml/l) or Dicofol 18.5% SC (2.0 ml/l) or Triazophos 40% EC (2.5 ml/l) is required to control other insects. The important diseases are damping off (*Pythium* sp., *Phytophthora* sp.), blight (*Phomopsis vexans*), little leaf etc. Seed treatment in hot water (52⁰C) for 30 minutes or with ceresin or dipping of seedlings with any copper fungicide is recommended for controlling damping off disease. The blight can also be checked by the same way. Little leaf is caused by virus with leaf hopper as the vector. To control this insect, parathion is found beneficial. Among the others root knot nematode is often found to attack. For effective control of nematode one have to follow appropriate crop rotation, use resistant varieties and spray fumigants like Nemagon.

3.2.4 Harvest and Post Harvest Management

The seed crop is thought to be mature when the fruit colour has started to fade. For example, the dark colour of a "black" fruit cultivar would start to become a bronze colour. Jayabarathi *et al,* (1990) opined that fruits harvested at the completely yellow stage had the highest seed yield (102.55 kg/ha), and harvesting fruits prior to or later than the full yellow stage resulted in lower seed yield and quality. Normally the fruits are harvested by cutting the peduncle, but for seed purpose the fruit is kept undisturbed in the plant to mature and ripen sufficiently and until an abscission layer develops between the fruit and calyx, causing the fruit to fall to the ground. A long brinjal can produce 800-1000 seeds; whereas, a round one can give upto 1000-1500 seeds. The seed yields of brinjal vary with different varieties or parents and production conditions. Generally, the standard of seed yield is between 600–800 kg/ha. The cost of hybrid seed production of brinjal is not as high as compared to other vegetables because each fruit contains a large number of seeds. The cost can be further reduced by the use of male sterile line in hybrid seed production. This advancement has made the exploitation of hybrid vigour in brinjal more economical.

Post harvest operations include extraction, drying and grading of seed. The extraction procedures may be two, namely, wet and dry. The wet method is more or less the same as in tomato. Seeds are extracted from the fleshy fruit in a similar fashion with only exception in that some extra water is added in the mixture for easy separation of seeds from the relatively dry fruit. In acid extraction, concentrated hydrochloric acid at 30 ml/kg of seeds is added in the pulped fruits with constant stirring for 20 minutes. For dry extraction of seeds, the fruit is left for drying further after picking. During the period of drying, the fruits shrievel and fading of colour begins. Seed is, then, extracted by hand. The method is mainly confined to small scale operations with plenty of labour. The drying of seed is done in the sun or in different driers. Seeds are conditioned to approximately 11% moisture content. But for vapour-proof storage the seed moisture content is reduced to 6%. The dry seeds of brinjal may, sometimes, require some milling for removing the fragments of dried fruit tissue. Grading is also necessary to separate unwanted small seeds from the lot and is done by floatation technique into floating and sinking batches. According to Selvaraj and Ramaswamy (1988), the seeds that sink is double in endosperm weight than that of the seeds that float and eventually the seeds that sink give better overall results. The average seed yield is 100-120 kg/ha, but good yield may reach 600-700 kg/ha Seed germination and vigour is better upto 15 years of storage, but decline rapidly thereafter. Selvaraj (1988) found highest germination and vigour in seeds treated with thiram and bavistin @ 1 g each per kg of seeds and stored in paper-aluminium foil-polythene-lined pouches.

3.3 Chillies/Pepper

Botanical name   : *Capsicum* sp.

Family   : Solanaceae

Chromosome No.   : 2n=24

Pepper may be of two types-sweet pepper (*Capsicum annuum* L.) and chilli pepper (*Capsicum frutescence* L.). Being a native of tropical America, both the types are widely grown throughout the world, the former being cultivated to a comparatively larger extent. In the temperate regions, it is grown as annual crop. The fruits of both the species are used fresh as well as processed for different purposes. The major chilli growing countries are India, Mexico, Japan, Nigeria, Thailand, Indonesia, China and Pakistan. India is the world's largest producer, consumer and exporter of chili peppers. Among which the city of Guntur in Andhra Pradesh produces 30% of all the chilies produced in India, and the state of Andhra Pradesh contributes to 75% of all the chilli exports from India. In India, Andhra Pradesh, Karnataka and Maharashtra account for almost three-forth of the country's net area and production. The seed production technique for both *Capsicum annuum* and *Capsicum frutescence* are more or less the same.

3.3.1 Botany

There are many cultivars differing from each other in shape and colour of the fruits, pungency and position of fruits. Bailey (1949) divided pepper into five groups based on fruit shape. *Cerasiformae* (cherry pepper, a pungent variety); *Conoides* (cone pepper, also pungent with conical or oblong cylindrical fruits); *Fasciculatum* (red cluster with fascicled fruits, red in colour and extremely pungent); *Longum* (long pepper, with drooping elongated pungent fruit); *Grossum* (bell or sweet pepper having large, puffy fruit with a depression at the base and usually furrowed sides. The fruit is red or yellow with a mild flavor. Based on taxonomic and genetic studies, Heiser and Smith (1953) included all the types and varieties mentioned above under *C. annuum* and listed the pungent variety Tabasco, together with some other uncommon varieties, as belonging to *frutescens*. Plants are herbaceous, dichotomous branches turn woody and brittle with age. Leaves simple, ovate to elongated. Flowers normally solitary but sometimes found as small cymes of leaf axils. Calyx five-lobed; corolla white and in five parts; five stamens separated but attached to the base of corolla; style one, longer than stamen; ovary mostly three-locular. Fruit is a berry with thin or thick pericarp according to the variety. Fruit setting percentage is the maximum during early anthesis which subsequently declines with the abortion of flowers (Khah and Passam, 1992). The 1000-seed weight of chilli is approximately 6g.

3.3.2 Environmental and isolation requirements

Pepper is basically a hot-weather crop and very much susceptible to frosts. The most productive crop thrives in a temperature range of 20-28 ^{0}C. A soil with pH ranging from 6.0 to 6.5 is optimum. The required ratio of NPK nutrition should be 1:1:1.5. Any excess of nitrogen is harmful as it delays the start of anthesis. High light intensities increase yield but reduces the 'capsaicin' content (the pungent principle) in the fruit.

The crop should not be rotated with other members of Solanaceae until a minimum of four years elapsed. Though essentially a self-pollinated crop, certain percentage of cross pollination occurs in pepper. So, an isolation distance of 400 m is recommended for foundation seed. Greater isolation distance is considered safer between distinct types (*e.g.*, pungent and sweet peppers).

3.3.3 Cultural requirements

Usually seedlings are produced in seedbeds. Seeds are also found to be sown directly in the field, in some parts of south-western America. According to Chauhan (1981), April to June is the best time of sowing (both in hills and plains), though the crops can be grown at any time of the year. About 1-1.5 kg seed can provide the required seedlings for one hectare of land. Seeds are often mixed with sand for easy sowing. The seeds are treated with *Trichoderma viride* @ 4 g / kg or *Pseudomonas fluorescens* @ 10 g/kg and sown in lines spaced at 10 cm in raised nursery beds and covered with sand. Watering with rose can has to be done daily. Drenching the nursery with Copper oxychloride @ 2.5 g/l of water at 15 days interval against damping off disease and application of Carbofuran 3 G at 10 g/sq.m at sowing is necessary. The nursery area is covered with 50% shade net and the sides are covered by using 40/50 mesh insect proof nylon net. For transplanting, 15-20 cm tall and 6-8 weeks old seedlings are found ideal. The planting distance should be 30-60 cm apart in the rows with 50-100 cm between rows.

Depending upon the variety, season and moisture retention capacity of the soil, the irrigations are to be given at 5-7 days interval, the first essential one being just after transplanting. Regular weeding is very much essential. To check the weed growth, 2-3 shallow hoeing during early stages of crop growth is beneficial. Application of Pendimethalin 1.0 kg a.i. / ha or Fluchloralin 1.0 kg a.i. / ha as pre-emergence herbicide followed by hand weeding once 30 days after planting is also found effective. The usual schedule of manuring and fertilization is as follows:

Crop	FYM(t/ha)	N(kg/ha)	P_2O_5(kg/ha)	K_2O(kg/ha)
Irrigated	25	45	80	40
Rainfed	25	30	50	25

Vanagamudi *et al* (1990) opined that for increased fruit and seed yield, the amount of nitrogen is to be enhanced upto 125-200 kg/ha. In areas of severe leaching loss, upto 60% of the nitrogen fixed as basal is better to be applied as top dressing after the establishment of plant but not after the start of anthesis. Spraying of Triacontanol @ 1.25 ml/l on 20, 40, 60 and 80th day of planting as well as NAA 10 ppm (10 mg/

l of water) on 60 and 90 days after planting is effective in increasing fruit set. Foliar spray of ZnSO$_4$ @ 0.5 per cent thrice at 10 days interval from 40 days after planting is beneficial. Roguing is done on the basis of plant and fruit characteristics as a whole, rather than the individual characteristics. There are three stages of roguing. The first is before anthesis, the second is after anthesis and the third is when the first fruit matures. Occasional off-colour plants can be rogued at the stage of fruit ripening.

Pepper is susceptible to some seed borne diseases like *Colletotrichum capsici* (anthracnose), *Phytophthora* sp. (blight, fruit rot) and *Pseudomonas solanacearum* (wilt). Infected plants are to be removed from the field during roguing. The crop is also attacked by mosaic, leaf curl, blossom end rot and die back. Among the major pests, thrips, aphids, mites and nematodes are important. Seed treatment with Captan/Thiram/Ceresan @ 2.5 g/kg of seeds may control fruit rot and die back. For reducing wilt incidence seed treatment with hot water at 50^{0}C for 30 minutes and application of neem cake is effective. Application of Furadon 3 G @ 30 kg/ha is effective against leaf curl. For controlling thrips and aphids, application of neem oil 1% or neem cake extract 5% or insecticides *viz.*, Monocrotophos @ 0.05% or Nuvacron @ 1.25 ml/l or Furadon 3 G @ 30 kg/ha may be beneficial. Mites and nematodes, on the other hand, can be controlled by spraying 0.02% methylparathion and aldicarb 10 G (20 kg/ha), respectively, during transplanting.

3.3.4 Harvest and Post Harvest Management

The fruits of sweet and chilli peppers are harvested after 50-60 days of planting or more (depending on varieties), when they are deep red colour and fully ripe. The fruits are picked, cut and seeds are separated mechanically. Fermentation is generally not practiced. The extracted seeds are dried to approximately 10% moisture content. But the safe moisture content for seed storage in vapour-proof condition is 4.5%. For seed storing, glass jars or polythene/plastic boxes are comparatively better than cloth bags or tin containers. The average yield of pepper seed may range from 40 to 60 kg/ha or even more, *i.e.*, 50 to 80 kg per hectare (Agarwal, 1980).

3.4 Onion

Botanical name   : *Allium cepa* L.

Family   : Alliaceae

Chromosome No. : 2n=16

Onion is one of the most popular and commercially important bulb vegetables grown in India, China, Japan, Russia, USA and several European and far Eastern countries. It comprises a major area in the world and used either for dry bulb or for green plants as salad. The probable origin of onion is Central Asia and the Mediterranean region. Onion bulbs are rich in minerals like P and Ca and carbohydrates. It is an important and indispensable item in every kitchen as condiment and vegetable. It is also used in soup, sauce, seasoned food as dehydrated bulb or powder and with many medicinal properties. As onion seeds are poor in keeping quality and lose viability within a year, it is essential to produce fresh seeds and use the same for bulb production.

3.4.1 Botany

The common onion (*Allium cepa* L.) produces bulbs in the first year and seed in the second year. So it is annual when grown for bulb and biennial when grown for seed. The leaves arise from crown stem with the outermost leaves enclosing the younger ones. The basal portion of the leaves encircle the stem and bulbs are formed. During the second year, the stem is elongated to form the flower stalk. Flowers are borne in simple umbels at the tip of a floral stem. Depending on the variety, the number of seedstalks per plant (with a height of 1 m or more) varies from 1 to 20 or even more and the number of flowers per umbel, on the other hand, varies from 50 to over 2000. The flowers are whitish with six stamens, three each in two whorls and a three-celled ovary with two ovules each. At the time of flower opening the style length is about 1 mm, which is not receptive until it elongates upto about 5 mm, 1 or 2 days after all of the anthers have dehisced. Flower opening continues for 2 weeks or more and the plants may bloom for over a month. Fruits (known as capsule) are 3-lobed and 3-celled, each containing 1 or 2 black seeds at maturity.

3.4.2 Environmental and isolation requirements

Onion requires cool weather during early stages and in early phases of growth of the seedstalk. Afterwards, bulb maturity and seed production during the second year is facilitated by a moderately high temperature and low atmospheric humidity. Onion favours deep friable loam soils (sandy and silty) retaining adequate amounts of moisture and with pH ranging from 6.0 to 6.8. Generally NPK requirements are in 1:2:2 ratio, with occasional topdressing of nitrogen, in some cases.

Stipulation on minimum rotation requirements is most essential not only to minimize the incidence of soil-borne diseases (*e.g.*, basal rot, pink root rot, etc.) but also to avoid nematode and certain other pests. Usually the isolation distance is kept as 800 m for foundation seed and 400 m for certified seed.

3.4.3 Cultural requirements

Onion seeds are produced by following two basic methods namely 'seed to seed' and 'bulb to seed' method. The first method is not so much usual. In this method, seeds are sown in late summer or early autumn and produced *in situ*. The seed rate and spacing to be maintained are 4-5 kg/ha and 70-100 cm (in rows), respectively. Sometimes, transplanting of seedlings from seed is also followed leading to flowering and seed production.

In the bulb to seed method, quality seed production is possible compared to the seed to seed method. Generally 'mother bulbs' are produced first and selected for good bulb characters during the first growing season. Then they are either stored or replanted immediately depending on the climate. The seed sowing time for growing mother bulbs varies from region to region. Seeds are sown in October and transplanted in November in north India. In hilly areas, seeds are sown in late February to May and transplanted in March to June. In addition to FYM, the NPK to be applied as per ICAR (New Delhi) recommendation is 50:50:100 (kg/ha). After a month of planting, 50 kg N/ha may be top dressed. According to Bhonde *et al* (1989), plants with a spacing of 40X30 cm and receiving 80 kg N/ha produced the highest seed yield (6.24 q/ha). Ground level irrigation is preferred to overhead irrigation with a view to reduce the incidence of some disease causing pathogens. One light irrigation just after planting is required followed by the other ones at 7-10 days interval. Weed-free cultivation is desirable to avoid any sort of competition causing hampered growth and development of the crop particularly in early stages. Two to three hoeing, not very deeper, are sufficient to check the weeds. Sometimes herbicidal options are also chosen.

Off-type plants and bulbs are rogued at four different stages. The first roguing is to be done before bulb maturity, the second is during sorting of the 'mother bulbs', the third is at planting of 'mother bulbs' and the fourth is done immediately prior to the start of flowering. In hybrid seed production, roguing of each of the inbreds used as parents is necessary. Regular checking and managing the incidence of pests (*e.g.*, onion thrips, mites, cutworms, etc.) and diseases (*e.g.*, damping off, purple blotch, stemphyllium blight, smut, neck rot, soft rot, black mould etc.) is among another important essentials. Thrips can be controlled effectively by applying phorate @ 10 kg/ha in the soil before transplanting. For controlling mites, infected bulbs should be exposed to sulphur @ 22kg/ha. Soil application of carbofuran @ 1 kg ai/ha at the time of planting may be used as preventive measure against cut worms. Seed treatment with thiram @ 2.5 g/kg seed is necessary alongwith soil drenching to control damping off. Spraying of copper oxychloride is the most effective and economical option in reducing purple blotch. To control stemphyllium blight a fortnightly spray of dithane M-45 @ 0.25% alongwith sticker triton is effective. Seed treatment with thiram @ 2 g/kg bulb and heating bulb at 40^0C for 8 hours before storage are effective against smut and neck rot, respectively. For reducing soft rot of cultivation of early varieties are recommended and field curing of bulbs for 10-12 days is necessary after harvest. Black mould attack can be minimized by dusting with $CaCO_3$ @ 12 kg/100kg at the cut end during harvesting.

3.4.4 Harvest and Post harvest management

Onion heads are hand harvested when the fruit opens and exposes the black seed. The ideal time of harvest is 45-60 days after the beginning of flowering in the field. In some areas, mechanical harvesting is followed but it can cause a significant seedless if not organized carefully. After cutting the umbels, however, should not exceed 15-20 cm. Otherwise, they should be turned regularly to avoid damage from overhealing or from fungal infection. Under humid condition, drying may be done in sheds. According to Brewster (2008), the drying temperature should not exceed 32^0C until the moisture content is less than 19%; 38^0C until the moisture content is less than 10% and 43^0C when less than 10%. Quick drying ensures minimum water absorption, thereby avoiding the possible danger of sprouting or reduction of viability.

After adequate drying, seeds are easily separated by rubbing in the hand carefully. Then they are threshed and subsequently cleaned either by hand flailing on small scale production or in threshing machines with carefully adjusted concaves. Further cleaning can be achieved by indent cylinders with gravity separator.

The seed should not be stored in bags or other containers until it is thoroughly dry. Depending upon variety, production technique and place of production, the average yield of onion seed is 500-1000 kg/ha. As the seed is considered as a relatively precious commodity, it is usually dried down to 6% moisture content and stored in vapour-proof condition.

3.5 Cabbage

Botanical name   : *Brassica oleracea* var. *capitata*

Family   : Cruciferae

Chromosome No. : 2n=18

Cabbage occupies a substantial area in India during winter and grown in UP, Karnataka, Maharashtra, Bihar, West Bengal, Punjab and Haryana. It is cultivated in 0.245 M ha with the total production of 5.617 M mt and average productivity of 22.9 mt/ha. Among these states West Bengal contributes 1.929 M mt of cabbage from 65,000 ha area with an average productivity of 29.6 mt/ha. Cabbage is used as salad, boiled vegetable and dehydrated vegetable as well as in cooked curries and pickles. Cabbage is rich in minerals and vitamins A, B1, B2 and C. Previously, the seeds of this crop are used to be imported from Europe and only during the Second World War attempts of cabbage seed production were undertaken when the supply was cut off (NCPAH, 2013).

3.5.1 Botany

Cabbage is basically biennial, but usually grown as an annual. The plant has fibrous root system. Plants are 40–60 cm tall in their first year at the mature vegetative stage, and 1.5–2.0 m tall when flowering in the second year. Cabbage is a temperate crop; thermoperiod is the most important factor for its flower induction. Flowers develop on the main stem. The inflorescence is an unbranched and indeterminate terminal raceme measuring 50–100 cm with flowers that are yellow or white. Each flower has four petals set in a perpendicular pattern, as well as four sepals, six stamens, and a superior ovary that is two-celled and contains a single stigma and style. Two of the six stamens have shorter filaments. The fruit is a silique that opens at maturity through dehiscence to reveal brown or black seeds that are small and round in shape.

Cross pollination is more common compared to self pollination. The main pollinating agents are the bees. The stigma remains receptive for about five days before and four days after anthesis. This facilitates "bud pollination technique"(artificial opening of flower bud and application of previously collected pollens to the exposed stigma) as a common method usually adopted by cabbage breeders. After pollination, 5 days are required for fertilization. The seeds are small, smooth and globular, 12-20 in numbers. About 225 g of seed (test weight 3.2 g) may be produced by an open-pollinated plant.

3.5.2 Environmental and isolation requirements

Cabbage thrives best in a relatively cool and moist climate. A well drained fertile soil not deficient in boron, manganese and molybdenum with a pH of 6.0-6.5 is ideal for cabbage. The optimum NPK ratio should be 1:2:2.

Between one cultivar to the other, a minimum isolation of 1000 m is necessary and the same between a type to the other should be at least 1500 m.

3.5.3 Cultural requirements

There are two methods for cabbage production:

(a) Head to seed method

(b) Seed to seed method

For stock seed production, the first method is followed since it gives a chance of proper inspection and roguing. But for the production of marketable seed, the second method is often employed.

(a) Head to seed method

The time of planting is so adjusted that a full head maturity is obtained just prior to winter (mainly in August or early September). Application of chlormequat @ 1.5 and 2.5 per cent applied during the full rosette stage has been reported to give better seed yields of 1.42 and 1.49 t/ha, respectively compared with 1.34 t/ha for control. When the head crop approaches maturity, off type plants are rogued. For storing the headplants in cellars, entire plants with roots on wire mesh arranged in tires are placed as close together as is practical for operation at 0^0C

and 90-95% RH. Before replanting in October-November, the top of each head is cross-cut and then planted in furrows at a depth of about 5 cm for producing seed in the next year. The spacing within and between rows are 45-90 cm and 100-200 cm, respectively. The cross-cuts are generally 2.5-5.0 cm deep made at right angle across each head only to facilitate normal development of the seed stalk.

(b) Seed to seed method:

The seeds are sown in beds from mid May (late varieties) to mid July (early varieties) and replanted at the end of July to early August at a spacing of 40-60 cm and 120-180 cm within and between rows respectively. Sambandamurthi and Sundaram (1989) opined that a low temperature (3-4^{0}C) over 2 months followed by a summer with small amount of well-distributed rainfall, are the preconditions for successful seed production in cabbage. During November-December, roguing of off-type plants is required. Sometimes cabbage needs adequate protection from extreme low temperature particularly during the winter months.

Both of the above two methods require some common cultural practices like staking, manuring watering, weeding and protection from pests and diseases. Staking is necessary to provide support for the developing flower stalks. For this, about 2 m long stakes are placed in upright position at 25 to 30 cm spacing along the row. Cabbage being a heavy feeder requires judicious fertilizer application particularly nitrogen, the amount of which may be ranged from 80-100 kg/ha under unirrigated condition to about 170 kg/ha under irrigated condition. If the soil is deficient in phosphorus and potassium, it is to be supplemented with the same as and when necessary. Thorough irrigation and weeding in proper times are extremely essential for cabbage. For controlling grassy weeds, herbicides like isopropylphenyl carbamate are found effective. For plant protection from pests like aphid (*Brevicoryne brassicae*), diamond black moth(*Plutella maculipennis*), cabbage worm(*Pieris rape*), pod weevil(*Centorynchus assimilis*) etc., and diseases like mosaic (virus), downey mildew (*Peronaspora parasitica*),black leaf spot (*Alternaria brassicola* and *A. brassicae*), clubrot (*plamodiophora* brassicae) etc. Suitable measures are to be taken. For controlling aphids, spraying of malathion @0.1% or monocrotophos @0.04% or nuvacron @0.05% is found wffwcti moreover, as the virus causing mosaic is transmitted by aphids, isolation of the seed field from the infected ones as well as control of aphid vectors are recommended. For controlling black leaf spot, seed treatment at 50⁰C for 30 minutes is sufficient. Clubrot can be managed through the use of benomyl.

Cole crops are susceptible to micro-nutrient deficiency. In cabbage boron and molybdenum deficiency has been reported by many workers. In case of boron deficient plants the stem becomes hollow and the plant growth is stunted which causes the condition known as 'whiptail'. This may be controlled by applying borax or sodium borate at the rate of 20 kg per hectare. In case of acute deficiency, spraying of 0.25 to 0.50 per cent solution of borax at the rate of 1 to 2 kg per hectare would give satisfactory control. Due to molybdenum deficiency young cabbage plants become chlorotic, the leaves become cupped and wither. Eventually the leaf dies and the growing plant also collapses and 'whiptail' develops. Application of 0.2 per cent Mo as foliar spray would give satisfactory control.

3.5.4 Harvest and Post harvest management

During harvesting of seeds, care should be taken for older pods to prevent shattering losses and the appropriate stage of maturity is judged by observing a significant proportion of pod turning yellow. Individual plants are usually cut and spread on floor to collect shattered seeds before they are threshed in a grain thresher or combine harvester. The seed yield ranges between 450 to 1000 kg/ha depending on crop genetics and other environmental factors.

The small, smooth and globular cabbage seeds are separated after cleaning in any standard mill utilizing screens and suctions. As the seeds are relatively brittle, slow cylinder speed is advisable. Light or shrievelled seeds can be removed by a gravity separator or with a spiral separator. Cabbage seed yield varies from 200 to 1000 kg/ha. The seed yield depends on prevailing temperature during the growing season, cool temperature during flowering and seed development results in higher seed yield. In India, Arya *et al* (1983) obtained 110.3 to 1203.8 kg/ha of cabbage seed yield with an average of 568 kg.ha.

3.6 Cauliflower

Botanical name   : *Brassica oleracea* var. *botrytis* L.

Family   : Cruciferae

Chromosome No.   : 2n=18

Cauliflower is an important vegetable in India and abroad. India produces 4.694 M mt of cauliflower per year from 0.256 M ha area with an average productivity of about 18.3 mt/ha. In West Bengal, the area under cauliflower is 57,000 ha with total production of 1.670 M mt and average productivity of 29.3 mt/ha. The major cauliflower producing states are Bihar, Uttar Pradesh, Orissa, West Bengal, Assam, Haryana and Maharashtra. The edible part of cauliflower is 'curd', which is made up of abortive flowers having short, fleshy and crowned stalks. It is intensively grown throughout India and having nutritive food value. Seed production of cauliflower is different from cabbage though the cultural practices are more or less similar. The major difference lies in the requirement of temperature. Cauliflower is more susceptible to freezing injury than cabbage (NCPAH, 2013).

3.6.1 Botany

Similar to cabbage, cauliflower is also biennial, but grows as an annual for curd. It may require low temperature treatment for flower induction in late varieties. However, the Asian varieties are of annual type and can flower and produce seeds under tropical conditions. The leaves of cauliflower are usually longer and narrower than cabbage. The inflorescence is smaller than cabbage and umbrella-like. After initiation of flowering more than 15 days are required for opening.

3.6.2 Environmental and isolation requirements

Similar to cabbage.

3.6.3 Cultural requirements

Seed production of cauliflower starts with sowing of seeds in early September and transplanting of seedlings in late October or early November. Depending on variety, the spacing differes, *viz.*, 100 cm X 75 cm and 120 cm X 90 cm for early and late varieties, respectively. The water supply should be uninterrupted throughout the growing period. However, excess moisture is harmful. Like cabbage, a crosscut is given in the card at maturity for better drainage of rainwater from that parts and allowing better aeration to the developing branches. For preventing any sort of decay due to cutting, spraying of Bordeaux is essential. Roguing of off type plant is necessary to bring uniformity in the variety taking due consideration on size, colour and compactness of the head. The insect-pests and diseases of cauliflower are similar to those of cabbage. Aphid is considered as the most serious pest. According to Raj and Kanwar (1990), endosulfan @ 0.05% , methyldemeton @ 0.05% and malathion @ 0.05% were effective against cauliflower aphid. Tate and

Cheah (1983) recommended captafol for root drenching at 0.1 g /200 ml to control clubrot disease. For preventing bacterial soft rot, Bordeaux mixture is found effective. Whiptail, another serious malady, can be managed by removing molybdenum deficiency.

3.6.4 Harvest and Post harvest management

Harvesting of cauliflower seeds require extreme carefulness for minimization of shattering losses and is usually done in September-October. Plants are cut with pruning shears while they are still yellow before placing them in an inverted position or wires stretc.hed across a Dutch barn and cured. After proper threshing, milling and cleaning, the seeds are collected. The average seed yield ranges from 150 to 200 kg/ha, depending on curd size. Liming is found to increase seed yield (by soil application in the form of paper mill sludge @ 250 kg/ha), but molybdenum has no effect in this regard (Panigrahi *et al.,* 1990).

3.7 Radish

Botanical name   : *Raphanus sativus*

Family   : Cruciferae

Chromosome No.   : 2n=18

Radish is a popular crop in both tropical and temperate regions. Radish originated probably in China. In India, it seems to have been cultivated from ancient times. It was popular among the ancient Egyptians and Greeks.

It is a good source of vitamin C and some minerals. It has some unique medicinal properties and also considered as an appetizer. Radish roots are eaten raw as salad or cooked as vegetable.

3.7.1 Botany

Banga (1976) described four present-day types of radish:

  i. *Raphanus sativus* var. *radiculata*—salad type, widely grown

 ii. *Raphanus sativus* var. *niger*—large rooted type

iii. *Raphanus sativus* var. *mougri*—pods type with edible leaves

 iv. *Raphanus sativus* var. *oleifera*—fodder type

All the above types show free cross pollination with each other and with *Raphanus raphanistrum* (wild radish). Bees and other insects are main pollinating agents of *Raphanus*. Except *Raphanus sativus* var. *niger*, which is biennial, all types are produced as annuals.

Both primary root as well as the hypocotyls produce the fleshy edible portion of radish root which is 2.5 to 10.0cm in length, oblate es of scarlet and crimson colour in case of annual varieties. Roots of biennial varieties may be 20 cm or more in length and black, purple or white in colour.

Inflorescence is a terminal raceme. Flowers are white, rose or lilac. Radish is highly cross-pollinated crop, pollination occurs primarily by honeybees. Flowers open during the day time from 8:00 A.M. onward. Dehiscence generally takes place at warmer temperature condition. Radish fruits differ from that of other crucifers in that it is not a siliqua but a true pod, about 2.5 to 7.5cm with a pithy interior. The colour of mature seed is initially yellowish, later turning reddish brown. Test weight is more (13.3g) compared to cabbage and cauliflower (about 3.2g).

3.7.2 Environmental and isolation requirements

Radish is a cool season crop. However, a wide climatic range is adaptable for it. For successful crop growth, fertile loam soil is desirable and it is advised to avoid light sands and heavy clays. A range of pH between 5.5 to 6.8 is optimum. The basal NPK ratio should be 1:3:4. The minimum isolation distance between crops are 1000 m and 1600 m for certified and foundation seed, respectively.

3.7.3 Cultural requirements

So far as seed production is concerned, radish cultivars available in India can be classified into three groups. Winter radish or Japanese radish (biennial) which produces seeds only in temperate hills of India. These cultivars require low temperature for flowering and are generally sown in the autumn i.e during the second fortnight of September. The second group includes summer radish of temperate regions e.g. White Icicle, Rapid Red White Tipped, Woods Long Frame, etc. These cultivars though very quick in root development behave just like winter radishes for seed production. In the hills the seeds of these cultivars can be produced both from autumn and spring sown crops. The autumn sown crops give higher yield and mature earlier than spring sown crops. The third group includes cultivars which produce seeds freely in the plains and can produce good seeds in the hills also. Generally, seeds of cultivars belonging to first two groups are produced in the hills only.

Seed production of radish can be possible by the following two methods:

(a) Root to seed method

(b) Seed to seed method

The former method is in vogue to produce high quality stock seed. Whereas, the majority of radish seed crops are produced by the later method, because it is simpler and less time consuming.

(a) Root to seed method

In early spring the seeds are sown by broadcasting or in rows with a seed rate of about 10 kg/ha. When the stecklings reach edible maturity, they are dugout within 3-4 weeks, graded and planted in another field with 100 cm between the rows and 15 cm within the rows. Some researchers opined about foliar spraying of GA_3 @ 50-75ppm, MH @ 25-50 ppm or CCC @ 250-500ppm during flowering stage. A spacing of 45 cm X 45 cm with 100 kg N/ha in radish cv. Pusa Rashmi produced the highest (6.5 q/ha) seed yield (Rawat and Singh, 1981). Care should be taken in roguing off type radish roots considering uniformity in shape, size and colour of roots.

(b) Seed to seed method

The sowing time is the same as in root to seed method. The seed rate is 3 kg/ha and row to row spacing is 70-90 cm with 2.5-5.0 cm distance within the rows. Careful and sufficient roguing at proper time is essential to raise seed to seed crops. Seeds sown in the second week of October showed highest seed yield (12.83 q/ha).

For both the above two methods some common cultural practices like manuring, watering, weed management and plant protection are essential. Application of 50 kg nitrogen with 25 kg phosphorus per hectare is found to give highest seed yield.

3.7.4 Harvest and Post-harvest management

Harvesting is usually done during August- September. Turning of pods into brown colour is the index of maturity. At this stage the plants are cut and forked into small piles for drying. Then threshing is done with bean threshers and thoroughly dried followed by cleaning. All the broken and sprouted seeds are removed. The seeds immediately after extraction are dried in the sun; otherwise they will lose their viability. Besides, high moisture content of seeds results in poor germination. The average seed yield is between 500 and 1000kg per hectare. Higher yields up to 1400 kg/ha can be obtained under favourable condition.

3.8 Cucurbits

Crop	Botanical name	Family	Chromosome No.
Cucumber	*Cucumis sativus* L.	Cucurbitaceae	2n=14
Muskmelon	*Cucumis melo* L.	Cucurbitaceae	2n=24
Watermelon	*Citrullus lanatus* Schrad.	Cucurbitaceae	2n=22
Pumpkin	*Cucurbita moschata* Poir	Cucurbitaceae	2n=40
Ridgegourd	*Luffa acutangula* Roxb.	Cucurbitaceae	2n=26
Spongegourd	*Luffa cyllindrica* Roem.	Cucurbitaceae	2n=22
Bittergourd	*Momordica charantia*	Cucurbitaceae	2n=22
Bottlegourd	*Lagenaria ciceraria* (Molina) Stendt	Cucurbitaceae	2n=22

Cucurbits form an important group of vegetable. Different crops are used in different ways, *viz.*, as salad (cucumber), for cooking (gourd), for pickling (cucumber), as dessert fruits (watermelon and muskmelon) or candied or preserved (ashgourd). Cucurbits are mostly seed propagated, but some are also propagated vegetatively like pointed gourd, chow chow and ivy gourd. This wide range of vegetables are grown basically in mixed cropping system and also in river beds.

Excepting few vegetatively propagated ones, the seed production in all cucurbitaceous vegetables are more or less similar to that produced for vegetables. The only difference is in allowing fruits to mature in plant itself for extracting seeds in full maturity instead of picking fruits at vegetative maturity.

3.8.1 Botany

The family cucurbitaceae consists of about 117 genera and 825 species out of which about 15 different species are being cultivated in India. The root system of all the economic cucurbits is extensive but shallow. Upon germination of the seed, the plants soon develop a strong taproot. The stems are branched (3 to 8), prostrate, trailing and usually angled in cross section. Fruits are essentially a berry, even though called a pepo, because of hard and tough rind (when completely mature) as in bottle gourd. The fruit peduncle is 5 to 8 angular. The edible portion is placentae in cucumber and watermelon, pericarp with very little mesocarp in pumpkin, while the whole fruit in bottle gourd. The seeds vary in size, shape, colour, the presence or absence of a margin and in the type of scar formed at the hilum. In general, each seed has a firm testa of several layers, a thin collapsed perisperm and endosperm and a large embryo. The embryo consists of two large, flat, leaf like cotyledons and a small radicle.

Flowering in cucurbits normally starts in about 40-45 days after sowing depending upon the weather condition. The sequence of flowering follows a set of pattern, namely (i) male phase: First few nodes bear only the staminate flowers, (ii) Mixed phase: both pistillate and staminate flowers appear in few nodes in the main axis and secondary branches in cycles and (iii) female phase: few nodes produce mostly the pistillate flowers. In a typical monoecious sex form of cucurbits the ratio of staminate and pistillate flowers may range from 25 to 30:1 to 15:1, the later condition is advantageous and economical, because consequently it results higher fruit set and yield. Generally high nitrogen, long days and high temperature promote greater number of staminate flowers.

Usually fruit set takes place early in the morning between 6:00 A.M. to 8:00 A.M. in crops like cucumber, pumpkin and watermelon etc. Monoecious condition in cucurbits imposes a situation conducive to cross pollination; however, a limited percentage (20-40%) of natural self-pollination takes place within the same plant. The andromonoecious condition favours a higher degree of natural self-pollination than in the monoecious condition. For maximum fruit set and seed yield, two bee colony per hectare would be required. a single fruit of a cucurbit contains a large number of seeds, hand pollination should be followed for quality seed production in cucurbits. In case of hand pollination male and female flower should be bagged before the day of anthesis. Afterwards, when anthesis take place, butter paper bags are opened and petals of male flower are removed and anther are gently rubbed on the stigma, then again female flowers are to be bagged for 2-3 days to avoid contamination by foreign pollen. After 3 to 4 days of pollination, bags can be removed.

3.8.2 Environmental and isolation requirements

Cucurbits are widely grown in the tropics and subtropics. Summer season is preferred over rainy season for raising seed crops, as seeds of some crops like bottle gourd or bitter gourd having thick seed coat do not dry properly in humid weather.

Seed production of watermelon is usually confined in areas where the ambient temperature does not fall significantly below 25⁰C. The crop needs an NPK ratio of 1: 1:1, but any excess in the amount of nitrogen is harmful. Muskmelon prefers a soil pH between 6.0-6.8 and NPK ratio of 1:1.5:2. Cucumber requires a soil pH of 6.5 with 1:2:2 of NPK ratio. In case of soils where leaching of nitrogen is excessive, the ratio of NPK should be 2:1:1. Other species of cucurbits are more tolerant to acid soil conditions and succeed on soils with a pH of 5.5-6.8. All the cucurbits, in general, respond to bulky organic base dressings of up to 25-30 t/ha.

For raising seed of cucurbitaceous crops, isolation is one of the basic requirements because of cross-pollination basically due to separate staminate and pistillate flowers in the same plant and entomophilous condition as well. Generally 500 to 1000 m isolation is found adequate for any cultivar of cucurbit species. Crops for basic seed production should be isolated by 1500 m from other species in this group. Complete homozygosity is not prevalent even in pure cultivars. In cross pollinated crops, genetic drift occurs during seed production if selection process (specially mass selection) is not undertaken. Again, because of being entomophilous cros, provision of adequate pollinator like bees are necessary.

The isolation distances of some important cucurbits are given below (Singh, 2001):

Crops	Isolatation Distance (m)	
	Foundation Seed	Certified Seed
Cucumber	800	400
Bittergourd	800	400
Bottlegourd	800	400
Ridgegourd	800	400
Spongegourd	800	400
Pointedgourd	800	400
Watermelon	800	400
Muskmelon	800	400
Pumpkin	800	400

3.8.3 Cultural requirements

The ideal sowing time for cucumber seeds are May-June and February-March for rainy season and summer season varieties, respectively. Watermelon seeds are sown in February in Northern India and from February to June in Central India. In South India, however, it is sown from mid-January to end of March. In India, seeds of summer squash are generally sown from January to April in the plains and October-November for early market production. The seeds of winter squash are usually sown from January to June in the plains. A seed rate of 3.5 to 4.5 kg/ha is sufficient for muskmelon; whereas for watermelon it is slightly higher (5.5-7.0 kg/ha). The seed rate of summer and winter squash are 8-10 and 5-7 kg/ha, respectively and the viability of seeds remain for about 4 years under good storage conditions. Seeds of cucumber are sown in rows or in pits 120-150 cm apart. For muskmelon, seeds are sown in ridges and furrows (150 cm wide with a distance of 60 cm between furrows) on both sides of the ridges with 90 cm between plants. Four to six seeds are sown at each hill followed by a light irrigation to promote germination. Watermelon seeds are sown in flat or raised beds on both sides or in pits with 3-4 seeds per pit.

Adequate fertilization, irrigation, weed control and plant protection are required for almost all seed crops. The hot season crops may require occasional watering compared to that of the rainy season crop. About 25-50 tonnes of manure per hectare should be incorporated in soil during final land preparation. NP and K fertilizers at an adequate ratio are to be ensured. Weeds, pests and diseases should be controlled by adopting suitable intercultural and plant protection practices. Irrigation should be stopped at the time of fruit ripening. The ICAR (Indian Council of Agricultural Research) recommendation for muskmelon and watermelon is 340 kg of ammonium sulphate, 540kg of single super phosphate and 136 kg of potassium sulphate per hectare. An application of 20g of lime per square metre has been found to be beneficial.

The principal cucumber pests are stripe and spotted beetles which can be controlled by applying 0.75% rotenone. Fruitfly (*Daucus cucurbitae*), red pumpkin beetle (*Raphidopalpa fovecollis*) and epilachna beetle (*E. emplicata*) are the major insect-pests of cucurbits in India including some other minor pests like aphids, plume moth, leaf eating caterpillar, blister beetle and mites. Downey mildew (*Pseudoperonospora cubensis*), powdery mildew(*Erysiphe polygoni*), leaf spot (*Cercospora* spp.), bacterial wilt (*Erwinia tracheiphila*), anthracnose(*Colletotrichum cucumerinum*), mosaic(virus) and root knot(nematodes) are the important diseases of cucurbits.

Most off-type plants are to be removed until the time of maturityand roguing should be carried out before flowering. The entire plant rather than the individual characters should be considered as the basic unit in the roguing operation. Muskmelon can be safely grown adjacent to cucumber, pumpkin, squash or watermelon since it will not cross with any of those crops.

3.8.4 Harvest and Post harvest management

Harvesting time is determined by observing specific maturity indices for different crops. In melon, seeds of all fruits cannot be extracted on one day; rather it may require 2-3 pickings. In cucumber, variation in flesh colour, external rind colour or spine colour should be critically taken care of and to be discarded accordingly. In case of all dry fruits like bottle-, sponge- and ridge gourds including summer squash, seeds are extracted when the fruits dry and seeds rattle inside the shell. The shells are broken for extraction of seeds and subsequent cleaning. In other cucurbits, where seeds are mixed with pulp, the seeds are scooped out at maturity with little attached pulp as in muskmelon, watermelon and cucumber etc. and collected in a barrel. Then seeds are rubbed with sand or ash and washed. Otherwise fermentation for 48 hours is done. Acid treatment is another rapid method of extraction of seed from pulp. 25-30 ml of hydrochloric acid acid for 5 kg pulp-containing seeds or 8-10 ml of commercial sulphuric acid can be used. This process may require 20-30 minutes. After thorough washing (to remove excess acid as well as to help floating ill-developed seeds), the seeds are dried in sun upto <8% moisture level before storing in containers in a cool room well protected from rats.

3.9 Cowpea

Botanical name   : *Vigna unguiculata* (L.) Walp.

Family   : Leguminosae

Chromosome No.   : 2n=22

Cowpea is probably the native of Central Africa. Nigeria is the world's leading cowpea-producing country. Other countries in Africa, for example, Ghana, Niger, Senegal, and Cameroon, are also significant producers. Outside Africa, the major production areas are Asia, Central America, and South America. Brazil is the world's second-leading producer of cowpea seed, producing 600,000 tonnes annually. The crop is grown grown all over India. It is cultivated both for vegetable and seed purposes. Green pods contain protein, carbohydrate, vitamins A, B, C and minerals like calcium and iron. Most of the commercial varieties are of bush type. Some of the important varieties of this vegetable are Pusa Phalguni, Pusa Barsati, Pusa Dofasli, Arka Garima, Pusa Suman and Pusa Rituraj.

3.9.1 Botany

It is an annual crop with white flower. Three subspecies of cowpea are cultivated in India, *viz.*, *Vigna unguiculata* sub sp. *unguiculata*, sub sp. *cylindrica* and sub sp. *sesquipedalis.*

Plants may be either erect, bushy with short branches or prostrate spreading in climbing forms.The leaflets are ovate to lanceolate, 5-18 cm long and 3-16 cm wide, entire or lobed and sub-tended by inconspicuous stipules. In the axil of each leaf there are three buds. Only the central bud normally expands to produce either a potentially indeterminate monopodial branch or a racemose inflorescence. Each axillary inflorescence is a compound raceme of several simple racemes carried on a grooved peduncle 5-60 cm. Each simple raceme has between six and twelve flower buds, but only the lower, first formed pair develops while the rest degenerate to form extra-floral nectarines between the paired flowers. The cleistogamous flowers are typically papilionaceous and are large, with a standard petal 2-3cm wide. Mature fruits vary widely in size, shape, colour and texture. Fruit lengths range between 12-20 cm and are straight, curved or coiled. They may lack pigmentation (tan or straw coloured) or have varying intensities of anthocyanin pigmentation from pink purple or almost black. They contain between six and twenty-one kidney shaped, oval or, rarely, almost spherical seed whose dimensions is within the range 5-12 mm and with individual weights between 50 mg and 340 mg.

3.9.2 Environmental and isolation requirements

Cowpea is a warm season crop and grown both in rainy and spring seasons. Crop cannot tolerate heavy rainfall particularly during flowering and pod set. It is a drought tolerant crop. Different varieties respond differently to temperature and day length. There are specific varieties for spring and rainy seasons. So varieties should be sown during their respective sowing time.

Crops can be grown on wide range of soil. Soil should have adequate drainage facilities. Cowpea does not prefer very acid soil.

Isolation distances of 50 and 25 m should be maintained respectively for production of foundation and certified seeds. All off-types and diseased plants should be eliminated at pre-flowering and flowering stages. Third roguing is done during pod maturity stage to receive late maturing plants. Cowpea should be isolated from Asparagus providing some distance.

3.9.3 Cultural requirements

Generally the crop is sown in January-February and May-June. But in the places of mild climate with moderate summer and winter, the crop can be grown round the year. Soil is made into a fine filth with 3-4 ploughings. Seeds are sown by broadcasting and drilling in rows 45-60 cm apart. Plants in rows are thinned at a distance of 10-15 cm in case of bushy verities. Seed rate is 20-25kg/ha. Before sowing, seeds should be inoculated with *Rhizobium* culture for quick modulation on roots.

FYM @ 20-25 ton/ha should be incorporated into soil at the time of final land preparation. About 25 kg N, 75 kg P_2O_5 and 60 kg K_2O/ha are required for the crop. Half of N along with full amount of P_2O_5 and K_2O should be applied into the soil before sowing. Rest N is applied into the soil during earthling up after third week of sowing.

Shallow cultivation is required during early stages of the crop. Crop covers the land very soon and kills the weeds by smothering. Irrigation is not required during rainy season. But summer crop needs irrigation at 4-5 days interval. Flowering and pod development period are the critical stages. So irrigation should be given during these stages it sufficient moisture is not available in soil.

To control powdering mildew spraying of wettable sulphur (3g/l) is done. Spraying of Dithane M 45 (2g/l) is done to control anthracnose disease. Aphid can be controlled by spraying the crop with 0.1% Malathion or 0.05% Phosphamidon and pod borer with 0.05% Endosulfan or 0.2% Carbaryl.

3.9.4 Harvest and Post harvest management

For seed purpose crops become ready for harvesting in 75-125 days. Maturity is determined by straw colour of pods. Dried pods are harvested or whole plants are harvested. Plants are dried after which they are threshed. Seeds are separated by winnowing. Clean seeds are dried till its moisture percentage reaches at 9. Seeds are packed in polythene coated bags and in cool and dry place. Seed yield is 10-15 quintal per hectare.

3.10 Indian bean

    Botanical name   : *Dolichos purpureus* L.

    Family   : Leguminosae

    Chromosome No. : 2n=22,24

It is native to Africa and it is cultivated throughout the tropics for food. are annual or short-lived perennial vines. The wild species is perennial. The thick stems can reach six meters in length. The leaves are made up of three pointed leaflets each up to 15 centimeters long. They may be hairy on the undersides. The inflorescence is made up of racemes of many flowers. Some cultivars have white flowers, and others may have purplish or blue. The fruit is a legume pod variable in shape, size, and colour. It is usually several centimeters long and bright purple to pale green. It contains up to four seeds. The seeds are white, brown, red, or black depending on the cultivar, sometimes with a white hilum. Wild plants have mottled seeds.

3.10.1 Botany

Indian bean or *Dolichos* bean is the native of India and mainly grown for its green pods while dry seeds are used as pulse. Two types of *Dolichos* bean *viz.*, *Dolichos purpureus* var. *lignosus* (field bean, annual type) and *Dolichos purpureus* var. *typicus* (garden bean, perennial type) are grown in India. The former is grown for tender pods, while the later is for dry seeds and used as pulse. Dark coloured mature seeds contain trypsin inhibitor which must be boiled before eating. Dry seeds contain 15-25% protein which is easily digestible.

Pusa Early Prolific, Arka Jay, Arka Vijay, Pusa Sem 2, Pusa Sem 3 are some important varieties of *Dolichos* bean.

3.10.2 Environmental and isolation requirements

Garden type prefers relatively cool season. Fruiting starts in the beginning of winter and continues upto early spring. It may be grown as a dryland crop. The crop can be cultivated on a wide range of soil with average fertility. Sandy loam, silty loam and clay loam soils are the best.

For seed production the crop should be isolated from field of other varieties. Maximum isolation distances of 50 m have to be provided for foundation and 25 m for certified seed production. Rogue out all off type plants, plants of other varieties, diseased plants and weeds through field inspection before flowering, at flowering and fruiting stages. Roguing should be done on the basis of plant type, foliage colour, pod shape, pod size and colour and maturity period of pod.

3.10.3 Cultural requirements

Land preparation is done by 3-4 ploughings. Sowing time is July-August. About 20-30 kg seed is required for sowing one hectare land. For quick nodulation on root, seed inoculation with *Rhizobium* sp. may be done. Pole types are planted in rows 1-1.5 m apart where the hills are spaced at a distance of 1 m.

About 20-25 tonnes of FYM is incorporated into the soil during final land preparation. Cop is fertilized with 20 kg N, 60 kg P_2O_5 and 60 kg K_2O per hectare. Half dose of BN and full of P_2O_5 and K_2O should be applied at the time of sowing. Rest of N is topdressed one month after sowing.

It is a hardy crop and commonly grown under rainfed condition. Shallow cultivation during early stages of crop growth is necessary until vines spread on field. Staking is done in pole-types. Light irrigation should be given according to the necessity of the crop. Flowering and pod development periods are the critical stages. So optimum moisture in soil should be maintained during those stages to reduce flower and fruit drop.

In anthracnose (*Colletotrichum lindemuthianum*) disease, dark brown or black sunken cankers surrounded by yellow margin appear on pods. To check the disease the crop should be sprayed with Carbendazim (0.2%). Powdery mildew disease is caused by the fungus *Erysiphe polygoni*. Whitish poedery spots first appear on leaves and later on stem and pods. Dusting of sulphur powder @ 20-25 kg/ha or spraying of 0.2% Karathane may render effective results against powdery mildew. Chlorosis, mottling, stunting and downward curling of leaves are the symptoms of common bean mosaic disease. This viral disease is transmitted by aphid. Resistant varieties like Kentucky Wonder, Contender should be grown in disease prone areas. Beside crop should be sprayed with 0.05% Malathion 50 EC to control insect vector.

Aphid (*Aphis craccivora*) suck the sap from leaves, stems, flowers and pods maybe controlled spraying the crop with 0.05% Phosphamidon or 0.1% Malathion. Young caterpillar of pod borer first feed on surface of the pod then bore into them. Endosulfan (0.05%) or Carbaryl (0.2) at 15 days interval may be used against this insect. Due to attack of bean thrips pods become silvery white in appearance. Spraying of 0.05% Hetahystox is effective in this case.

3.10.4 Harvest and Post harvest management

Crop is harvested when the majority of the pods are fully ripe and remaining has turned yellow. Harvesting should be done before pods become over matured to avoid shattering and seed injury. It is better to harvest the pods during. After harvesting the crop is dried for 7-10 days. Then the produce is threshed by beating with a stick, by bullock or by a thresher. Seeds should not be injured mechanically at the time of threshing. After threshing, seeds are cleaned and dried to keep the moisture level below 9%. Seeds are kept in polythene coated bags and stored in cool and dry place. Seed yield is 15-18 quintal per hectare.

3.11 Sweet Potato

Botanical name : *Ipomoea batatas* (L.)Poir.

Family : Convolvulaceae

Chromosome No. : 2n=90 (Hexaploid)

It is an important starchy food and usually taken after boiling, baking or frying. Tubers are also used for canning, dehydrating and flour manufacture. It is the cheapest source of calorie having enormous variation in tuber size and colour. The tubers are also used to prepare sweet, jam, alcohol and syrup. It can be cultivated even without irrigation ensuring a good profit per unit area. In India, sweet potato is basically grown in Bihar, Orissa, UP, Maharashtra, Karnataka and MP. The origin of this crop is South America.

The seed production of sweet potato means production of vines as well as tubers.

3.11.1 Botany

The sweet potato [*Ipomoea batatas* (L.)Poir.] Belongs to the family convolvulaceae. It is a herbaceous perennial, but cultivated as annual. Stem is a vine with milky juice, prostrate and twinning. It produces roots at the joint. Triangular leaves are borne on the stem the colours of which are green to purple. Root tubers develop on nodes of vine which rests on the soil by the process of secondary thickening of underground adventitious roots. The inflorescence of cymose bearing single or few flowers. The five-lobed corollas are funnel-shaped, purple in colour. Stammers are five in number, unequal in length and attached near the corolla base. The pistil a 2cm long style with 2-lobed stigma. Seeds are borne on capsule containing upto 4-black coloured seeds but one or two develop.

3.11.2 Environmental and isolation requirements

The crop prefers moderately warm climate during the growing period of four months and a temperature range between 21.1-26.7^{0}C. It cannot stand even a slight frost, but favours plenty of sunshine and a well distributed moderate rainfall. Flowering is stimulated by short days, long nights and high temperature. High rainfall and long photoperiod promote vine growth and reduce tuber yield. The best soil for sweet potato is a well-drained sandy loam with clay subsoil. Saline as well as alkaline condition can also be tolerated by the crop. A soil pH between 5.2 to 6.7 is appropriate for sweet potato.

3.11.3 Cultural requirements

From freshly harvested vines of mature crops, the cuttings are taken for direct planting. For getting good yield, selected tubers are to be planted in the nursery and propagated by "sprouts". The selected tubers should be healthy, medium-sized and free from disease-pest and bruises. Planting in a well prepared manured nursery having good drainage is done 3 months ahead of planting in the main field. The row spacing of tubers in the nursery bed is 60 x 25 cm. The beds are irrigated as per requirement. After about 40-45 days the sprouts are cut and planted in a secondary nursery for further growth at 60 x 25 cm spacing. After another 45 days, on reaching sufficient length, the cuttings are made from the nursery vines and planted in the main field at 60 x 30 cm or 60 x 45 cm or 45 x 45 cm spacing. Selection of tip or middle portion of the vine helps increasing number and yield of tubers. The vines are to be kept in shade for two days before planting to ensure early rooting. The old vines are the living place of insect. Before planting the vines are soaked in Monocroptophos @ 1 ml/l of water to reduce the infestation of insect. In the Gangetic alluvium of West Bengal, the highest tuber yield was recorded by planting vine cuttings in September-October. The vines are planted in ridge and furrows or in beds. In some areas the vines are buried inside the soil in the middle and one node at each end is kept exposed. Sometimes vertical planting is practiced. For a hectare of land, about 80,000 vines are required. For sweet potato nutrition, nitrogen and potassium are more important compared to phosphorus. An experiment in West Bengal revealed that the maximum economic yield of marketable tuber was obtained with 90 kg each of nitrogen and potassium per hectare. Phosphorus @20-25 kg/ha is found to be sufficient. Application of nitrogen and potassium, should be in two splits where irrigation facility is available. The most critical stage of moisture supply is 40 days after planting. Earthing up of soil is helpful at this stage (35-40 days after planting) to suppress weeds.

Incidence of disease is very less in sweet potato. However, foot rot is found in some areas. Seed treatment with fungicide *viz.*, Benlate. Bavistin or Captan @ 1g/l of water is found effective before planting vines in the nursery. Cultivation in the same land is to be avoided year after year. Other diseases are stem rot, leaf spot, black rot, scurf including soft rot, ring rot or collar rot at storage. Suitable fungicides alongwith crop rotation, disease-free planting materials and clean cultivation are essential. The most important and common pest of sweet potation India is *Cylus formicarius* (weevil). The incidence of this pest increases during summer starting from February-March. Spraying of @ 2 ml/l of water at an interval of 3 weeks reduces the attack. Recently killing the male insect through pheromone trap is becoming popular. Use of resistant cultivars is also effective.

3.11.4 Harvest and Post harvest management

Generally the crop is harvested when the leaves of vine become yellowish at fourth month after planting of vines and the soil cracks. Before harvest, the vines are cut, then the tubers are dug up by using spade or fork. An irrigation 4-6 days before harvesting is helpful in case of soil compaction. The yield may range from 35-40 t/ha in irrigated and 8-10 t/ha under rainfed conditions. In addition to this, 10-25 tonnes of vines are produced per hectare to be utilized as fodder or planting materials.

After harvesting, tubers are kept for healing for 4-5 days. This is curing. Then sorting is done by removing all cut, bruised, damaged, deformed and weevil infested tubers. The selected tubers and packed in bags are stored.

3.12 Palak

Botanical name : *Beta vulgaris* var. *bengalensis* L.

Family : Chinopodiaceae

Chromosome No. : 2n=18

It is one of the most common leafy vegetables of tropical and subtropical region in India. The succulent leaves and stems excellent nutritious diet. The probable origin of palak is Indo-China region. It is mostly grown up in UP, WB, Punjab, Haryana, Delhi, MP, Bihar, Maharashtra and Gujarat. The most popular Indian varieties are All Green, Jobner Green, Banerjee's Jiant, Pusa Palak, Pusa Jyoti, Pusa Harit, etc. However, this crop is not so much popular in South India.

3.12.1 Botany

For the edible leaf production it is grown as an annual, but the same is biennial for seed production. Plants produce succulent tender edible leaves in rosette form initially which results in bolting after about 75 days of sowing at elongated stems. The inflorescence is racemose with bracts of cymose type and emerges either in the axil of leaf or from a terminal bud from the main as well as lateral shoots. Flowers are small, sessile, bisexual, perfect and are borne in the axil of leaf in a group of two to three. Stamens are five, opposite to perianth lobes on a fleshy disc within unilocular perigynous ovary. The crop is highly cross-pollinated. Wind pollination is the general rule. The process of flower-bud development takes 35 days to come in bloom. The anthesis period is between 7.0 AM to 5.0 PM which reaches its maximum during 11.0 AM to 1.0 PM. Anther dehiscence takes place between 8.20 AM to 6.30 PM which reaches its maximum during 12.30 PM to 2.30 PM. Stigma receptivity begins 8 hours before anthesis, reaches the peak just after anthesis and continues for another 10 hours. The fruit is actually a seedball containing 2-3 seeds.

3.12.2 Environmental and isolation requirements

The crop favours mild climate. It can withstand frost and warm weather, but too much temperature leads to early bolting. The soil pH should range between 6.0 to 6.8. the NPK ratio should be 2:1:2 applied during final land preparation. Susceptibility to boron deficiency is found with this crop. Hence, application of borax @ 1.5-2.0 kg/ha is recommended.

It is a cross-pollinated crop and readily crosses with beetroot, swiss-chard and pigweed. But it does not cross with spinach (*Spinacea oleracea* L.). A minimum of about 1600 m isolation distance is generally given in palak for foundation seed production and 1000 m for certified seed production.

3.12.3 Cultural requirements

Palak is generally sown thrice in India in the plains, *viz.*, a) in early spring, b) in early rains and c) during September to November as main crop. However, it can be grown throughout the year in areas having optimum environmental conditions. The normal seed rate is 25-30 kg/ha. For hastening germination, seeds are sometimes soaked overnight. The sowing methods comprises of either broadcasting or line sowing in rows 20 cm apart. The recommended dose of inorganic fertilizer is 50 kg nitrogen, 60 kg phosphorus and 60 kg potassium per hectare as topdressing after cutting. The spring-summer season crop needs frequent watering at 6-7 days interval, whereas that of the autumn-winter crop is at 10-15 days interval. The rainy season crops need very less or no irrigation. Proper interculture like loosening of soil, weeding and rouging is necessary. Usually, after taking 3-4 cuttings, the crop is left for seed production but it is not desirable as removal of leaves reduces the seed yield. Roguing comprises of careful removal of off-types and early bolters. The first roguing is done at pre-flowering stage based on foliage characters followed by subsequent roguings based on early bolters and off-type plants.

3.12.4 Harvest and Post harvest management

Palak seeds mature in 150-180 days. After complete ripening of the fruit of the whole crop, harvesting is done. The plants are dried, and seeds are threshed by beating with sticks. Cleaning of seeds is done by winnowing and drying to 9% moisture level is performed before storage (in low humidity and temperature). The average yield of seed is 600 kg/ha.

3.13 Elephant Foot Yam or Ol

Botanical name   : *Amorphophallus campanulatus* Blume.

Family   : Araceae

Chromosome No. : 2n=26

'Ol' is a very popular vegetable in tropical and subtropical regions. The crop is grown commercially in India, China, Malayasia and Ceylon. In India it is generally cultivated in Andhra Pradesh, Bihar, Gujarat, Maharashtra and West Bengal. 'Ol' is a cheap source of carbohydrate and rich in protein, minerals as well as vitamins (A and B). apart from its wide use as vegetables, it is used for preparing curries and pickles. It has some medicinal uses also.

3.13.1 Botany

It is herbaceous, either annual or perennial. Leaves are simple and petiolate. The flower, known as spathe, is funnel- or bell-shaped at the base, springing from the corm with disagreeable odour. Two main cultivars, *viz.*, Santragachhi and Kovvur, are popular in eastern and southern parts of India. The average yields of these two varieties are 60 t/ha and 100 t/ha, respectively, depending on the size of the planting material.

3.13.2 Environmental and isolation requirements

The crop does not prefer extremes of temperature and demands a well-distributed rainfall during its growth period. During corm development, it requires a low temperature and dry weather. Water logging is harmful for the crop. Well-drained, well-aerated loam or sandy-loam soil is considered ideal for corm growth. It can be cultivated in hills and even in red lateritic soils with sufficient decomposed compost and light soil application in the pits.

3.13.3 Cultural requirements

The ideal time of planting is March-April in irrigated condition. Otherwise, the same is at the end of monsoon. Yield will be less when seed sowing is late. The propagating material is the mother corms or daughter corms. The offsets which grow out of the mother corms are known as daughter corm or buds. Big-sized mother corms are cut into pieces and used for planting. Considering the value of seed, usually seeds of 500 g to 1 kg weight are suitable irrespective of source. Generally the spacing depends on seed size. Seeds of size 1 kg each are spaced at 75 x 75 cm, whereas the same of 1.5-2 kg size are spaced at 100 x 100 cm or even more. Small corms are more effective than the cut ones because they yield 40% more, germinate early and never rot. At CTCRI, Trivandram, the maximum as well as economic corm yield was obtained by using 1 kg of corm as planting material (CTCRI,1983). It has been found that closer spacing increased the per hectare yield but decreased average corm yield. Planting is done in flat beds or ridges by making pits of size 50 cm x 50 cm x 30 cm. In each pits 2-3 kg cowdung or 250 g cake, 15 g urea, 18 g SSP and 18 g MOP is mixed well alongwith soil and the seed is placed. Then some more soil is used to cover the seed completely. To increase germinability, waterholding capacity and to reduce weeds, sometimes decomposed paddy straw or compost is given over each pit particularly in the drought prone areas. Irrigation is applied at 7-10 days interval from sowing of seeds. The seeds need 40-50 days to germinate. In that way generally 8-10 irrigations are needed till the onset of monsoon. After germination and before unfolding of leaves (40-50 days after sowing), topdressing is needed @ 15 g urea and 18 g MOP per pit. Care is to be taken during topdressing of fertilizers so that no harm of roots occur to prevent foot rot. Earthing up is done for better bulking of the corm and good drainage. The crop needs very less crop protection. To prevent foot rot, Rhizolex @ 5 kg/ha or thiride and Vitavax @ 2.5 kg/ha or Furadon @ 5 kg/ha is ato be applied 2-3 times.

To produce whole seed of 'Ol', generally 100-150 g sized pieces are made from big-sized corms (with a portion of head in each) and spaced at 30 cm x30 cm. from these seeds, about 70-75% plants will come out. The cut pieces (of size 100-150 g) are soaked in fresh cowdung water and dried or soaked in thiourea @ 300 mg/l of water for 6 hours and dried to increase germination capability and produce whole corm.

3.13.4 Harvest and Post harvest management

At the end of the season generally $750 \ g^{-1}$ kg sized corms are produced per plant. Th required fertilizers are comparatively lesser than that for ordinary cultivation. The seeds are kept for 3-4 months at cool room over paddy straw. The duration of the crop is 7-10 months. The maturity index is yellowing and drooping of leaves. The yield of corm varies depending on weight of seed corm used at planting.

3.14 Okra or bhindi

    Botanical name   : *Abelmoschus esculentus* L. Moench

    Family   : Malvaceae

    Chromosome No. : 2n=130

Okra is cultivated throughout the tropical and sub-tropical regions alongwith the warmer parts of the temperate regions. Okra plants are grown commercially in many countries such as India, Japan, Turkey, Iran, Western Africa, Yugoslavia, Bangladesh, Afghanistan, Pakistan, Myanmar, Malaysia, Thailand, India, Brazil, Ethiopia, Cyprus and in the Southern United States It has a good potential as foreign exchange earning crop accounting for approximately 60 per cent of the export of vegetables. It is cultivated in 0.349 M ha area with the production of 3.344 M mt and productivity of 9.6 mt/ha. The major okra producing states are Uttar Pradesh, Bihar, Orissa, West Bengal, Andhra Pradesh and Karnataka. In West Bengal, 0.662 M mt of Okra is produced from 58,400 ha with an average productivity of 11.4 mt/ha. The crop is also used in paper industry as well as for the extraction of fibre. Okra seeds contain about 20% protein and 20% oil. Okra seed can be dried, and the dried seeds are a nutritious material that can be used to prepare vegetable curds, or roasted and ground to be used as coffee additive or substitute (Moekchantuk and Kumar 2004). Okra seed oil is an acceptable feedstock for biodiesel production (Anwar *et al*, 2010).

3.14.1 Botany

The height of okra plants may be upto two meters. The plants produce large white to yellow flowers which develop into ridged pentagonal pods. The seed pods are 3 – 10 inches long, tapering, usually with ribs down its length. These tender, unripe seed pods are used as a vegetable, and have a unique texture and sweet flavour.

3.14.2 Environmental and isolation requirements

Bhindi is a warm season vegetable crop and requires a long warm growing season. Warm humid tropical conditions are ideal for luxurious growth and high yield of okra. It grows best within a temperature range of 24-27°C and is highly tolerant to high temperature and drought condition. But, the crop is highly susceptible to frost injury as severe frost causes damages to the pods. Temperature below 12^0C is detrimental to the crop and seeds fail to germinate when temperature falls below 20°C. But, the crop can be successfully grown in rainy season even in heavy rainfall area.

It can be cultivated in a wide range of soils from sandy loam to clayey loam. However, loose, friable, well drained loamy and sandy loam soils rich in organic matter are ideal for its growth. It also gives good yield in heavy soils with good drainage. But the crop cannot tolerate excessive moisture or poorly aerated soils. A pH range of 6.0-

6.8 is considered as optimum. Alkaline, saline soils and soils with poor drainage are not good for this crop.

3.14.3 Cultural requirements

The lands are prepared thoroughly by mechanical means or with the use of animal-drawn implements. Okra being direct seeded requires 10 kg seed/ha for sowing during June - August and February. Seeds are soaked in water overnight before planting for faster and uniform germination. Air drying of the seeds before sowing is necessary.Seed treatment with *Tricoderma viride* @ 4 g/kg or *Pseudomonas fluorescens* @ 10 g/ kg of seeds and again with 400 g of *Azospirillum* using starch as adhesive and drying for 20 minutes in shade is beneficial. Three seeds are sown per hill at 30 cm apart and then thin to 2 plants per hill after 10days.The required spacing is 45 x 30 cm. to ensure uniform seed germination irrigation immediately after planting is recommended. After two weeks of planting, weak and diseased seedlings are to be thinned out to maintain one healthy plant per hill. Rate of fertilization depends on soil analysis, but in its absence, 10 gm complete fertilizer (14-14-14) is to be applied. At 30 and 45 days thereafter, side dress 15 g of a mixture of 2 parts Urea (46-0-0) and 1 part Muriate of Potash (0-0-60) per hill. During dry months, furrow irrigation is done every 7 days. Water is critical at planting, after emergence, during the vegetative stage and at flowering and fruiting development.Tthe plants are to be kept weed free within the first month. Hand-weeding is advisable especially around the base of the plants. The most common pest of okra is leaf hopper. For managing leaf hopper, the seed treatment is essential with imidacloprid 48% FS or 70% WS @ 7 g/kg or Thiamethoxam 70% WS @ 2.8 g/kg of seed. Spraying of any one of Imidacloprid 70% WG 0.7 g /10 l or Thiamethoxam 25%WG 1.0 g /10 l or Azadirachtin 5% Neem Extract or Dimethoate 30% EC 2.0 ml/l or Malathion 50% EC 1.25 ml/l or Oxydemeton – Methyl 25% EC 1.6 ml/l Quinalphos 25% EC 1.0 ml/l is advisable. Another important pest is fruit borer. For this, integrated pest management *viz.*, setting up pheromone trap @ 12/ha, collection and destruction of affected fruit, release egg parasite of *Trichogramma chilonis* @ 1.0 lakh/ha, release 1st instar larvae of green lace wing predator *Chrysoperla carnea* @ 10,000/ha, dusting of carbaryl 10% DP @ 25 kg/ha, spraying of *Bacillus thuringiensis* @ 2 g/l is recommended. Spraying of any one of the insecticides *viz.*, Azadirachtin 0.03% WSP (300 ppm) 5.0 g/l or Emamectin benzoate 5% SG 3.0 g /10 l or Phosalone 35% EC 1.5 ml/l or Pyridalyl 10% EC 1.0 ml/l or Quinalphos 20% AF 1.5 ml/l or Quinalphos 25% EC 8.0 ml /10 l is also proved beneficial. For diseases like Yellow vein mosaic virus systemic insecticides like Methyl demeton or Dimethoate @2 ml/l to kill the insect vector, whitefly are to be sprayed. And for powdery mildew dusting of Sulphur 25 kg/ha or spraying of Dinocap 2 ml/l or Tridemorph 0.5 ml/l or Carbendazim 1 g/l or Wettable sulphur 2 g/l or Triademephon 0. 5g/l immediately after noticing the disease along with repeat spray after 15 days is necessary.

3.14.4 Harvest and Post harvest management

Seeds mature about 110-120 days from emergence.The basal and apical pods do not mature all at the same time so that about 3-4 primings may be needed. Pods turn leathery brown in colour at seed maturity. Matured pods are put in canvas matting and sundried for 2-3 days or until pods become brittle.Pods are then carefully threshed to extract the seeds and clean using air-screen cleaner or winnow.All unfilled seeds are removed keeping only the healthy seeds.Then the clean seeds are gradually dried under the sun for 4-5 days to lower the moisture before being packed for storage. For home use, the seeds are packed in a thick plastic or paper envelopes and they are placed in large aluminum cans or large-mouth jars lined at the bottom with charcoal, lime or silica gel.The seeds are then placed in a cool, dry place after proper sealing. For large volume, the seeds are packed in thick plastic bags or aluminum-lined packets and sealed tightly. All seeds are kept in a cool and dry storage area for an enhanced shelf-life.

The seed yield is greatly influenced by the variety, season, location and management practices Under normal conditions the seed yield of 10-12 q/ha can be obtained. However, in certain pockets of Gujarat and Andhra Pradesh the higher seed yield of 18-20q/ha have been achieved.

Chapter 4

Seed Production (Spice Crops)

4.1 Turmeric

Botanical name   : *Curcuma longa*

Family   : Zingiberaceae

Chromosome No. : 2n=64

Turmeric has been used in India for thousands of years. It is an important spice crop of India and regarded throughout the world. Apart from being inevitable as spice in our day-do-day kitchen, it has several industrial and medicinal use. It is a major part of Ayurvedic medicine.It was first used as a dye and then later for its possible medicinal properties. It is one of the key ingredients in many Asian dishes. Indian traditional medicine called Ayurveda, has recommended turmeric in food for its potential medicinal value, which is a topic of active research.Turmeric grows wild in the forests of South and Southeast Asia. India is the largest producer and exporter of turmeric in the world, the major importer countries being UAE, Saudi Arabia, Iran, Japan, USA and UK.

4.1.1 Botany

Turmeric is an erect perennial herb. The pseudostem is 50-120cm high with short stem and long, broad, green lanceolate, acuminate leaves. The rhizome is thick, brownish with secondary and tertiary branches all forming a dense clump. The flowers are pale-yellow, borne in spike in the leaf axils. Calyx are short, toothed and split. Corolla is tubular at base and cup-shaped at upper portion. Ovary is inferior and trilocular. Flowers do not set viable seed. Flowering duration is June to October. Flowers open between 6.0-6.30 AM. Pollination occurs through insects.

4.1.2 Environmental and isolation requirements

Turmeric thrives warm and humid climate with well-distributed annual rainfall of about 1000-2000m and 20-30⁰C temperature for normal growth and rhizome development. Both high and low temperature has adverse effects. It may be grown in sandy loam to heavy clay soils at 5-7.5 pH. For seed production, the crop is kept isolated by 5 m between varieties or other crops. This is to avoid the chance of mixing rhizomes during harvest.

4.1.3 Cultural requirements

Turmeric fingers or rhizomes are used for propagation. Both ridge and furrow or flat-bed method of planting is followed. The former one is preferred in areas where water stagnation is frequent. The planting time ranges from April to July, the peak being in May-June. For ridge and furrow planting a spacing of 45-60 cm (row to row) and 15-20 cm (plant to plant) is maintained. Whereas, in case of flat bed method 30 x15 cm spacing is the best. Planting depth for both the method is 3-5 cm. the optimum rate of planting materials is 2500 kg/ha of rhizomes. Before planting, the materials are treated with 0.25% Dithane M45 for 30 minutes. Usually 20-25 tones of FYM along with NPK @ 120:60:60 kg is applied per hectare. Nitrogen is applied in three splits; once at planting and the others at 3- and 4- months after planting, respectively. Full dose of phosphorus and half of potassium is to be applied before planting and the rest half of potassium is applied 4-months after planting. Top dressing of fertilizer is to be followed by slight earthing up as it helps mixing of fertilizers, gives support to the plants and provides sufficient aeration for rhizome development. Watering at frequent interval is necessary depending on soil types and climatic conditions. Mulching twice, *viz.*, just after planting and two month after planting with green/ dry leaves, sugarcane trash, dhaincha, paddy straw etc., is beneficial for soil moisture rentation, reduction of weed growth and sprouting of rhizome. Generally about 12-15 tones of green or 5-6 tones of dry leaves are required for mulching of one hectare land. Frequent hand-weeding application of chemical herbicide like oxyflurofen @ 0.15kg/ha is effective as pre-emergence application. Necessary measures are to be taken against shoot borer (*Dichocrosis punctiferalis*), scale insect (*Aspidiotus hartii*), lacewing bug (*Stephanitis typicus*), etc., by spraying phosphamidon @ 0.05%, dipping rhizomes in dimethoate solution @ 0.05% for 5 minutes and spraying malathion @ 0.1%, respectively. Among the diseases rhizome soft rot (*Pythium aphanidermatum*) is the most important one which can be controlled by spraying Bordeaux mixture (5:5:50) and planting healthy rhizomes or raised beds with necessary crop rotation.

4.1.4 Harvest and Post harvest management

Seed crop of turmeric is harvested at full maturity stage at 7-9 months after planting depending on varieties. Early harvesting causes poor yield, reduced keeping quality and increased physiological loss in weight. Deep ploughing is encouraged to collect clumps without any damage. Then the rhizomes are cleaned to remove adhering soil and root. After this, fingers are separated from mother rhizomes, both of which are suitable for seed purpose. Then the fingers are dried and cured for immediate marketing or for temporary storage.

4.2 Ginger

Botanical name	:	*Zingiber officinale* L.
Family	:	Zingiberaceae
Chromosome No.	:	2n=22

Ginger is another important rhizomatous spice of India after turmeric. Ginger is cultivated in India, China, Japan, Indonesia, Australia, Nigeria and West Indies. India is the largest producer and consumer of ginger in the world. In India, ginger is produced in Orissa, Kerala, Karnataka, Arunachal Pradesh, West Bengal, Sikkim and Madhya Pradesh. Kerala is the largest ginger producing state, accounting for about 33 per cent of the total production in India. Cochin ginger (NUGC) and Calicut ginger (NUBK) are the popular Indian ginger varieties in the world market. Except Jammu and Kashmir, it is grown in almost all states of India. Fresh ginger has its wide use as spice. It is also used in the preparation of bread confectionery, sauces, pickles, etc. Ginger oil and oleoresin is famous throughout the world. Ginger is popular for its unique medicinal properties.

4.2.1 Botany

Ginger is herbaceous, perennial and about 30-100 cm high. Leaves, 8-12 in number, are having long sheaths forming pseudostems. The rhizome is fleshy, hard, palmately branched, pale yellow and covered with small scales with fine fibrous roots. Inflorescence is cylindrical spike arising from the rootstock. Flowers are bisexual. Calyx and corolla are three-lobed, tubular and three each in number. Stamens are three, ovary is trilocular. Formation of fruit occurs very rarely. If at all produced, it is capsular and small. The flowering duration is between October to December. Anthesis and dehiscence is simultaneous at 3PM.

4.2.2 Environmental and isolation requirements

Ginger prefers warm and humid climate with well-distributed moderate rainfall from planting to sprouting of rhizomes (ranging 1500-3000 mm annually) and a day temperature range between 28-35⁰C. A dry spell of about one month prior to harvesting is effective. Medium loam soils rich in humus are ideal for ginger, though it can be grown in a variety of soil. The crop cannot stand waterlogging. For avoiding disease incidence, especially the rhizome rot, drainage is most essential and crop rotation is to be followed.

The minimum isolation distance is maintained at 5 m, similar to that of turmeric.

4.2.3 Cultural requirements

Rhizome and their parts (also known as sett or bit) having about 15-20g weight with at least one sound bud are used for propagation. The time of planting starts from April and continues upto July. Both flat and raised bed planting methods are followed depending on rainfall and drainage. The optimum seed rate is about 12-18 quintals of rhizome (15-20g each) per hectare. They are planted at a spacing of 20 x 15 cm the nutritional requirement of ginger is very high. For better development of rhizome, the soil should be provided with heavy amount of organic matter. Usually 15 tonnes of FYM alongwith NPK @ 120:60:60 kg per hectare is to be supplied. Full amount of phosphorus and half of potassium may be applied at the time of planting. Half nitrogen is applied 40-45 days after planting and the remaining half nitrogen and half potassium are applied 80-90 days thereafter. For avoidance of rhizome rot, neem cake @ 2 tonnes per hectare as basal is found useful.

Watering is to be done as and when necessary. Usually the majority of the crop being rainfed requires less water. But when there is no rain particularly during germination and developmental stages, irrigation is necessary. A light irrigation 5-6 days prior to harvesting facilitates easy harvesting of rhizomes. Weed control is a must for the crop. Hand weeding as well as herbicide application *viz.*, Simazine, 2-4 D, etc. is effective. Mulching in ginger beds by sugarcane trashes, dry leaves, grasses etc. is necessary to prevent soil erosion, retain soil moisture and enhance germination. Extreme care is to be taken during plant protection for removing diseases *viz.*, leaf spot, soft rot, bacterial wilt, storage rot, etc. and pests *viz.*, shoot borer, scale, leaf roller, nematode, etc. by applying necessary chemicals, clean cultivation, crop rotation, resistant cultivars, etc.

4.2.4 Harvest and Post harvest management

After 8-9 months of planting, ginger is ready for harvesting depending on the variety, requirement of use, agroclimate and management practices. Care is to be taken during the harvest so as to avoid all possible injury to the rhizomes. Under ideal situation, 20-25 tonnes of seed rhizomes are obtained from a hectare of land.

After harvesting, all the rhizomes are allowed to remain as such for 24 hours in the field. Then the soils are cleaned and tops as well as the root re separated by 2-3 times. Then the rhizomes are dried in shade for few days. After drying, sorting is done. Then they are stored properly.

4.3 Coriander

Botanical name : *Coriandrum sativum* L.

Family : Apiaceae

Chromosome No. : 2n=22

Coriander is a native of Mediterranean region. Plants and seeds have a pleasant aromatic odour. Its young plants are used for the preparation of chutneys, sauces and salads. Leaves are used for flavouring curries and soups. Seeds are used as condiment in pickling, seasoning, sausages and bakery products. Leaves and seeds are rich in vitamins A, B_2, B_3 and C. Seeds are carminative, diuretic, tonic, stomachic and retrigerant. Oil extracted from coriander seed contain dextrolinalool used for flavouring beverages, pickles, sweets, etc. India exported coriander to the tune of 2631.43 tonnes during June 2013. Major export destination countries were UAE, USA, Nepal, UK, Australia, Malaysia, etc. (Agriwatch, 2013).

Some important varieties of coriander are Pant Haritima, Swathi, Sadhana, Gujarat Coriander-1, Gujarat Coriander-2, Gwalior No. 5365, Merrocan, Rajendra Swati, UD 20, CO 1, CO 2 and CO 3.

4.3.1 Botany

Coriander is a herbaceous annual and sex is andromonoecious. The leaves are variable in shape, broadly lobed at the base of the plant, and slender and feathery higher on the flowering stems. White or pinkish flowers are born on terminal umbels. Each umbel comprises of several umbellets. Peripheral flowers of an umbel are generally hermaphrodite. Fruits are globular, dry schizocarp 3–5 mm in diameter, yellow brown in colour, globular, ribbed and consist of two mericarps.

4.3.2 Environmental and isolation requirements

It is a tropical crop and grown successfully during *rabi* season for seed production. Crop is susceptible to frost during flowering and seed formation stages. Heavy rains alongwith cloudy weather during these two stages is favourable for occurrence of diseases and pests. Cool and dry weather is essential during grain formation stage.

Loamy soil is the best for its cultivation. Soil should be rich in organic matter coriander can be grown as rainfed crop in heavy soils where moisture retentive capacity is more. Saline or alkaline soils are not suitable for cultivation of coriander.

Isolation distances are 800 m for foundation and 400 m for certified seed production. Seed production field should be rogued at pre-flowering, flowering and maturity stages. During pre-flowering stage all off types and volunteer plants should be removed from the field. Removal of all late flowering and diseased plants during flowering stage and all late maturing plants during maturity stage are done.

4.3.3 Cultural requirements

Thorough preparation of land is done with 3-4 deep ploughings followed by harrowing and plankings. A pre-sowing irrigation if required, is given to achieve uniform germination. Coriander requires 15 tonnes of FYM, 60 kg N, 40 kg P_2O_5 and 30 kg K_2O per hectare. FYM is applied during land preparation. Half of total quantity of N and entire amount of P_2O_5 and K_2O are applied before sowing. Remaining N is applied in two equal splits at 30 and 60 days after sowing.

In plains, coriander seeds are sown in June-July as kharif crop and in October-November as rabi crop. Sowing time in hill is during March-April. Both broadcasting and line sowing methods are followed for sowing of seeds. Seed rate is 10-15 kg/ha for irrigated conditions and it is 25-30 kg/ha when the crop is grown under rainfed condition. Before sowing seeds are splitted into two halves by rubbing and soaked in water for 12-24 hours to enhance germination percentage. Before sowing, seeds should be treated with Thiram @2.5 g/kg of seed to check attack of diseases. Spacing adopted for the crop is 25-30 cm x 15 cm.

Hoeing and weeding are done 30 days after sowing. Thinning operation should be carried out simultaneously during this time. Depending on the moisture availability in soil, irrigation may be given at an interval of 10-12 days.

Yield and quality of seed are reduced due to attack of powdery mildew (*Erysiphe polygoni*) disease. Spraying of 0.1% Karathane at 15 days interval or 0.1% Carbendazim at maturity stage is effective to control the disease. Wilt disease is caused by fungi *Fusarium oxysporum* and *Fusarium coriandrii*. Seed treatment, deep summer ploughing, crop rotation and use of some tolerant varieties namely, UD 20, Sadhana and Swathi are the possible control measures against wilt disease. Dark brown spots appear on stem and leaves in blight (*Alternaria poonensis*) disease. Spraying with 0.2% Carbendazim or 0.2% Mancozeb render effective protection. In stem gall (*Protomyces macroporus*) disease, blisters appear on leaf and stem which deform the seeds. This disease can be controlled by seed treatment with Thiram, crop rotation and using tolerant varieties like Rajendra Swati and UD 14. Aphid (*Hyadaphis coriandri*) is a serious pest of coriander. Its incidence is more during late flowering stage. The insect can be controlled by spraying of 0.03% Dimethoate, 0.05% Phosphamidon or 0.03% Malathion.

4.3.4 Harvest and Post harvest management

Generally crops become ready for harvest in 90-120 days after sowing. When colour of grains turns from green to yellow, indicates maturity. Crops are harvested when more than 50% of grains obtain yellow brown colour. Harvesting during this stage improve luster of grains. Delay in harvesting generally leads to shattering of seeds. Harvesting is done by uprooting or cutting the plants by sickle. Harvested plants are stacked in partial shade for 4-5 days. Then dried plants are threshed on clean floor by beating with a stick. Seeds are separated by winnowing. Clean seeds are dried in sun so that moisture level in seed is maintained at 10%. Dried seeds are bagged in gunny bags lined with polythene sheet and stored in a cool, dry and well ventilated place. Seed yield is 10-12 quintal per hectare.

4.4 Fenugreek

Botanical name	:	*Trigonella foenum-graecum* L. [Common methi]
		Trigonella corniculata L. [Kasuri methi]
Family	:	Apiaceae
Subfamily	:	Fabaceae
Chromosome No.	:	2n=16

Fenugreek is the native of south-eastern Europe. Major fenugreek-producing countries are India, Iran, Nepal, Bangladesh, Pakistan, Argentina, Egypt, France, Spain, Turkey, Morocco and China. The largest producer of fenugreek in the world is India, where the major fenugreek-producing states are Rajasthan, Gujarat, Uttarakhand, Uttar Pradesh, Madhya Pradesh, Maharashtra, Haryana, and Punjab. Rajasthan produces the lion's share of India's production, accounting for over 80% of the nation's total fenugreek output.

Green leaves of fenugreek are rich source of protein, vitamins A and C and minerals (calcium and iron). Seeds contain *diosgenin* which is used in the synthesis of sex hormone. Seeds have many medicinal properties and used in the treatment of diarrhoea, diabetes and many ailments. Green leaves are used as vegetable and seeds as condiment.

Some of the improved varieties of this crop are Pusa Early Bunching, Lam Sel. 1, Prabha (NLM), Pusa Kasuri Selection, Methi No. 47 and Methi No. 14.

1.4.1 Botany

Fenugreek is an erect annual with tri-foliate leaves, divided into toothed leaflets. It produces solitary or paired yellow-white flowers tinged with violet. The seeds are brownish, about 1/8 inch long, oblong, rhomboidal, with a deep furrow dividing them into two unequal lobes. They are contained, ten to twenty together, in long, narrow, sickle-like pods. The crop is self-pollinated and flowers hermaphrodite. Common methi is quick growing in habit and produces white flowers; but Kasuri methi is slow growing with bright orange yellow flowers.

4.4.2 Environmental and isolation requirements

Fenugreek is a cool season crop and can tolerate frost also. It is capable of growing in hot climate. Crop prefers moderate or low rainfall. Poor seed yield is obtained when the crop is grown in shady places. Weather should be dry during seed maturity period. Rains, high humidity and cloudy weather are harmful for seed production. Comparatively cooler climate is required for Kasuri methi.

Crop can grow in wide range of soil. But clay loam soil best suited for its cultivation. Soil pH should be 6-7. Land should be ploughed for 3-4 times so that soils are brought to a fine tilth.

Isolation distances are 50 and 25 cm respectively for foundation and certified seed production. All off types and diseased plants should be removed from seed production field on the basis of growth habit, leaf characters, flower colour and maturity. Roguing is performed at three stages of crop growth namely, before planting, during flowering and fruiting and during maturity.

4.4.3 Cultural requirements

For seed production of common methi, optimum time of sowing is September to middle of November in plains. Kasuri methi is sown in November. In hills both the types are sown during March-April. Sowing is done by broadcasting or by line sowing. The later method is better as it facilitates intercultural operations. Seed rate is 20-25 kg/ha. Spacing adopted for line sowing is 25-30 cm x 10 cm. higher seed yield can be achieved if seeds are inoculated with *Rhizobium meliloti*.

About 15 tonnes of FYM should be incorporated into the soil during final land preparation. Besides, the crop needs 40 kg N and 50-60 kg each of P_2O_5 and K_2O. Entire doses of P_2O_5 and K_2O and half of N are applied as basal. Remaining N is top dressed in two equal splits *viz.,* during flowering and fruit development stages.

At early stage of crop growth shallow cultivation is required to remove weeds. Thinning of plants is done to maintain proper spacing. A light irrigation should be given after sowing if sufficient soil moisture is not available. Subsequent irrigations are given at 10-15 days interval depending on soil conditions. Irrigation is essential during vegetative and pod development stages when moisture level in soil is not optimum.

Young plants are attacked by damping off (*Rhizoctonia solani*) disease. Drenching the soil with 0.1% Carbendazim can effectively control the disease. Powdery mildew (*Erysiphe polygoni*) disease appears late in season and it can be protected by spraying with 0.05% Karathane or 0.2% sulphur. Yellow spots on upper surface of leaf and cottony grayish white mycelium of lower surface are the common symptoms of downy mildew (*Peronospora trigonella*) disease. To control this disease copper fungicide can be used. Sometimes aphids also attack fenugreek plants. Spraying of 0.03% Dimethoate or 0.05% Malathion is effective against this insect.

4.4.4 Harvest and Post harvest management

Seed crop requires 120-160 days to harvest. But it depends on variety and sowing time. During harvesting time pods become yellow. Harvesting is done by cutting the plants at ground level. They are dried in sun on open floor. Seeds are separated by beating with a stick. Winnowing is done to clean the seeds. Seeds are then spread on open floor till moisture percentage reaches at 7-8. Seeds should be stored in a moisture proof polythene coated bags. Seed yield is 12-15 q/ha.

4.5 Coconut

Botanical name   : *Cocos nucifera* L.

Family   : Palmae

Chromosome No.   : 2n=32

Coconut, regarded as *Kalpa Vriksha* (Tree of Heaven), is a wonder palm in the world and almost all its parts are useful to manknd. It has been originated in south east Asia. The major coconut growing countries are in Asia, Oceania, West Indies, Central and South America, East and West Africa. Phillippines leads the world in area. Coconut production plays an important pole in the national economy of India. According to figures published in December 2009 by the Food and Agriculture Organization of the United Nations, it is the world's third largest producer of coconuts, producing 10,894,000 tonnes in 2009.

Traditional areas of coconut cultivation are the states of Kerala (45.22%), Tamil Nadu (26.56%), Karnataka (10.85%), Andhra Pradesh (8.93%) and also Goa, Orissa, West Bengal, Puducherry, Maharashtra and the island territories of Lakshadweep and Andaman and Nicobar. Four southern states put together account for 92% of the total production in the country

4.5.1 Botany

The genus *Cocos* is a monotype which contains the only species *Cocos nucifera* Linn. It is a tall palm of height 12-24 m. the trunk is stout raised from a swollen base surrounded by a mass of adventitious roots. It has an adventitious of root system typical of monocots producing numerous thick roots from the base of the stem almost throughout its life. Leaves are large, pinnate, borne on crown. Branched inflorescence enclosed in a sheath collectively known as the spadix which is 1.2-1.8 m long, stout, erect, straw or orange coloured, androgynous, simply branched. Branches (spikes) bear one or more female flowers between two male flowers towards bases and several males above. Male flowers are numerous, small; female flowers are larger and fewer than male. Ovary is tricarpic, usually one-ovuled. Fruits large, 20-30 cm long, subglobose, three sided, one seeded drupe. Outer pericarp is thick and fibrous, inner endocarp or shell is very hard with three basal pores representing the remains of 3 carpels. The thin testa cohering to the endocarp is lined with white albuminous endosperm (meat), with a little sweet fluid.

4.5.2 Environmental and isolation requirements

Coconut is a tropical plant and highly adaptive to a variety of environments. It is cultivated in altitudes ranging from 600-900 m. The optimum total rainfall per annum is between 1300 and 2300 mm. severe drought at flowering period is harmful. The ideal mean annual temperature is about 27^0C with a diurnal range of 6-7^{0}C. Coconut can be grown on heavier soils provided they are well drained. Otherwise, it prefers light soil. This palm prefers a soil pH ranging from 5.2 to 8.0.

4.5.3 Cultural requirements

Selection of mother palm is important in coconut. Mother palms should be of 25-60 years of age with at least 30-32 fully opened leaves on crown. Shape of crown should be spherical or semispherical. Petiole length and stalk of the bunch should be short and strong in nature. Bearing habit should be regular with at least 80 nuts annually during the last 5 years. The nuts should be medium sized, round to oblong (1200g in weight with dried husk). Drooping crown, dis-shaped nut, proximity of cattle shed and compost pit near the palm must be avoided. The seed nut collection time from the selected mother palm may vary from region to region. Usually the ideal time is from February to May having 11-12 months aged full mature nut. The presence of water in nut is judged by shaking the nuts and getting a clear metallic sound on tapping. The seednuts after collection are stored in open shade for about a month till the husk becomes dry to facilitate speedy and maximum germination. Then they are arranged on the floor of a shed over 7-8 cm thick layer of dry sand with their stalk-end up and covered with sand to prevent drying of nut water. Five layer of nut can be arranged one over the other. During summer, sprinkling of water is needed to prevent the drying of nut water.

Well-drained, coarse-textured soil in the vicinity of assured water source is desirable. The beds are prepared with a height of 10-15 cm from ground level to avoid water stagnation. As a precaution against white grub and termites, the soil is to be treated with HCH (10%) @ 60 kg/ha or Chlordane 5% dust @ 120 kg/ha or soil drenching with Chlorpyriphos (5 ml/l of water). Usually the beds are made of 2 m width and of convenient length with 60-75 cm space in between two beds for better supervision, irrigation, interculture and drainage. The seed nuts are sown at a spacing of 40 cm x 30 cm in 20-25 cm deep trenches during May-June. Before sowing it is better to soak the seeds in water to facilitate easy and early germination. The nuts may be placed horizontally or vertically. The advantage of vertical planting is on account of its convenience in transport. On the other hand, horizontal planting has been critically found advantageous as feeding the developing seedling by the endosperm is ensured by horizontal planting will be in close proximity with the developing embryo thereby providing more favourable condition for germination. During planting, the widest of the three segments is to be placed uppermost in horizontal planting to ensure early and higher germination percentage as well as vigorous seedling with thicker collar girth. The seednuts should be buried with soil and covered with thick layer of sand to prevent termite attack.

As a management requirement, irrigation, fertilization, weeding and plant protection is vital. Irrigation should be given at regular interval during dry season to ensure soil moisture as per the requirement of the seedlings. Fertilizers in the form of NPK @ 20:20:40 kg/ha is to be applied in three splits during December, February and April. Weeds are readily removed from the seedbeds. During hot and dry periods, coconut leaf mulch is found to be effective for early and better germination, good seedling growth and better establishment. Moreover, shading should be done immediately after the monsoon ends, especially when the nursery is raised in the open space. Careful inspection is a must for detecting the incidence of pests and diseases. As a measure to control scales and mealy bugs spraying of Dimethoate @ 0.05% on the undersurface of the leaves is effective. To check mites, it is necessary to spray Dicofol @ 0.05% from beneath the leaves. Fungal diseases like leaf spots and bud rot are checked by spraying 1% Bordeaux mixture or 0.3% Indofil-M 45 on both sides of leaves, removal and burning of severely infected leaves, *etc.*

The nuts start germination usually in 11-12 weeks after planting and continue upto 5 months. The criteria of seedling selection are early germination, early splitting of leaves into leaflets, short and thick leafstalks, healthy and robust appearance, minimum 6 leaves and collar girth of 10 cm at one year of age, freedom from pests and disease incidence, etc.

4.5.4 Harvest and Post harvest management

Harvesting of coconut is done with three main considerations like *copra*, coconut oil and tender coconut water. Besides the other considerations are Desiccated coconut (DC), coconut cream, coconut milk and spray dried coconut milk powder, coconut cake, coconut toddy, coconut shell. Wood based products, leaves and pith are also of interest. However, when seed production is the major objective mature and healthy dry coconut seed is to be harvested and stored properly till it is planted.

Commonly cultivated varieties / cultivars of coconut are attacked by various insect pests in store. Among these ham beetle, *Necrobia rufipes* and saw toothed grain beetle, *Oryzaphilus surinamensis* are of major importance, which can cause more than 15% loss to *copra* when stored for more than six months. Following precautions are to be taken for the safe storage of *copra* for more than three months:

- Dry the produce to four per cent moisture content.
- Avoid heap storage, which causes maximum damage.
- Store *copra* in netted polythene bags or gunny bags. (KissanKerala, 2013).

4.6 Arecanut

Botanical name	:	*Areca catechu* L.
Family	:	Palmae
Chromosome No.	:	2n=32

Arecanut or betel nut is an important tropical palm and form a popular masticatory in India and west Asia. Chewing pieces of arecanut alone or in combination with betel leaves is in practice from historical times. Production of arecanut in the world was about 10.33 lakh tones from an area of 8.29 lakh hectares in 2009-10. India ranks first in terms of both area (47%) and production (47%) of arecanut. The other countries which produce arecanut in the world are Bangladesh (21% in area and 9% in production), China (6% in area and 20% in production) and Indonesia (16% in area and 6% production). It is also cultivated in Myanmar and Thailand on a smaller scale. The world productivity of arecanut stood at 1.21 tonnes/ha. Indian productivity is also on par with the world productivity (1.27 tonnes/ha), (DGCIS, Kolkata). The main countries where arecanut is exported from India are Nepal, UK, Singapore, Maldives, Saudi Arabia, Thailand, Australia, USA, etc. in different years (DGCIS).

4.5.1 Botany

Arecanut is a plant with unbranched slender stem and dense crown. It is a cross-pollinated species. The spadix is enclosed in a boat-shaped spathe. Male and female flowers are separate. Female flowers are confined to the territory rachis and to the distal end of the secondaries. The male flowers are produced on filiform branches, which arise below and beyond the female flowers. The male phase begins on an average about 4 days after the spadix has emerged from the spathe and the phase lasts for about 26 days. On the other hand, the female phase lasts only for 3 days after the completion of male phase.

4.5.2 Environmental and isolation requirements

Since arecanut is cultivated in a variety of soil and climatic conditions, it is difficult to formulate uniform agronomical practice suitable for all the situations alike. In NE India, arecanut is mostly grown on palms, since winter temperature will be harmful for the crop at higher elevation. The ideal temperature range is 14^0C-36^0C. Extremes of temperature and wide diurnal variations are not conducive for the healthy growth of the palms. It can withstand very low (750 mm) as well as very high (4500 mm) rainfall per annum.

4.5.3 Seedling production methodology

Arecanut is propagated exclusively through seeds. As the plant is perennial, utmost care is needed right from the beginning to avoid any sort of difficulties in future thereby declining the growth and yield potential of the crop. Usually there are a few principal stages in the selection and raising of arecanut seedling for better growth and yield as follows:

(a) **Selection of mother palms :** Genetically superior mother palm possessing characters of high heritability are selected. The basic criteria of selection are age, earliness regularity of bearing, stabilized yield, high percentage of fruitset, freedom from pest and diseases and presence of 8-9 healthy leaves on the crown with shorter internodes. Generally the plants which produce at least 4 bunches per annum with a minimum yield of 250 nuts per bunch are to be selected.

(b) **Selection of seednuts :** The major factors considered for seednut selection from selected mother palm are: position of nuts in bunch, weight of nut (>35 g) within a bunch, maturity and floating habit of nut (vertical). The selected bunches are to be lowered to the ground carefully by using a coir rope to minimize damage of nut.

(c) **Storing of seednuts :** The nuts after harvest are smeared with cowdung slurry and kept under shade for 2-3 days to enhance germination. In some areas, farmers dry the seednuts in open sun for 1-2 days. Then they are stored in plastic bag, moist saw dust or moist sphagnum moss to prolong the storage period without affecting the percentage of germination. Arecanut seeds alongwith moist sand and 0.2% KH_2PO_4 (Seed:Sand ratio, 1:3) recorded high germination (85%) even after 4 months of storage when kept in polybag (350 gauge) and stored in cool energy chamber.

(d) **Nursery techniques :** Seedlings are raised in primary as well as secondary nursery beds. Bold whole nuts give better germination rate than half-husked or fully husked nuts. The nuts are to be sown in soil or sand immediately after harvest with regular irrigation. Nuts are spaced in rows at 5-6 cm with stalk ends up preferably at the end of winter. Arrangement of partial shade is necessary in the nursery with periodical irrigation, weeding, mulching and nutrition. To ensure easy transport and better handling, seednuts can be sown in polybags (25x 15 cm, 150 gauge) with potting mixture of soil:FYM:sand at 7:3:2 ratio. Germination commences in about 40 days after sowing. After 3-4 months, seedlings having 2-3 leaves are transplanted to the secondary nursery at a spacing of 30x30 cm during early monsoon. Necessary measures to control pests and diseases at nursery are to be taken.

(e) **Selection of seedlings -** Normally 1-1.5 year's old seedlings are transplanted in the main field by uprooting them with a ball of earth adhering to the roots. The seedlings with five or more leaves having short and stout stature are preferred. A scientific method for seedling selection is counting the number of leaves at the time of planting, multiplying the number by 40 and then subtracting the height of the plant. For example, a value of 140 cm can be obtained for a seedling having 6 leaves and 100 cm height [(6x40)-100=140]. If such values of plantable seedlings range from 60-150, then the seedlings having higher value may only be selected, discarding those having lower values. For quality seedlings of areca cv. Mangala using a selection index of 60 based on number of leaves and seedling height is recommended for planting since they possess short stature, wide collar girth and more leaves.

4.5.4 Harvest and Post harvest management

Mature nuts are harvested 8-9 months after flowering when the entire kernel has become hard amd hasks have turned bright yellow to orange when mature. To ensure maturity the lightest nut is cracked open on an inflorescence (cluster). If not mature, then the inflorscence is left intact only to be checked again at a later date. It is to be ensured, however, that the cluster is harvested before letting the fruit drop because once in contact with the soil the soil the nuts begin to germinate and probability of disease and pest infestation rises.

Cleaning of harvested nuts is usually done manually. The dirt, stalk end and other unwanted materials are removed by hand and the nuts are then rinsed or washed with plain water, after which they are dried in the sun and stored. The stage of harvesting depends on the type of produce to be prepared for the market. The most popular trade type of arecanut is the dried nut known as chali or kotapak. Fully ripe nuts – about 8–9 months old fruits having yellow to orange red colour – are best suited for the above purpose. Ripe fruits are dried in the sun for 35 to 40 days on a dry levelled ground. For drying and dehusking, fruits are sometimes cut longitudinally into two halves and sun dried for about 10 days, after which the kernels are scooped out and given a final drying.

Grading is done according to the purpose and use. Nuts with a thin husk and with an average weight of above 35 g are considered to be best grade and they fetc.h a good price. Good grade nuts are sorted out manually. Uniform, mature nuts, free from surface cracking, sticky husk and fungal or insect infestation, naturally command a good price.

Chapter 5

Seed Production
(Flowers and Ornamental Crops)

Flowers are high value horticultural commodities used in various ways in domestic and social activities as well as in industries (*viz.,* essential oils, dry flowers, natural dye extraction etc.). The enhancement in per capita income and urbanization has led to increased demand for flowers. Floriculture in India is growing at the rate of 7-10% per annum. The area under flowers is concentrated mostly in Andhra Pradesh, Karnataka, Maharashtra, Tamil Nadu, and West Bengal. A provisional estimate of National Horticulture Board puts the area under flower crops at 1,91,000 hectare with a production of 10,31,000 MT of loose flowers and 69,027 lakh cut flowers during 2010-11.

Good quality planting material (seeds, bulbs, etc.) is a basic need of a grower. But production of healthy and disease free planting material is a difficult task and required lot of experience, planning and management. The quality refers to genuine and diseased free material. Good quality planting material will boost productivity. Viability and good germination are of paramount importance in case of seed propagated ornamental crops (Marigold, Aster, Zinnia, Hollyhock etc.). In ornamental crops very little have been achieved in the production of genuine planting material free from the diseases. In recent years several multinational and industrial houses have entered into floriculture business and are producing quality plant material. Apart from private industries the intensive research conducted at premier research institutes like IIHR, IARI, NBRI and other SAUs had led to development and release of high yielding and disease free varieties. The planting materials of those elite varieties are supplied from these institutions in small quantities to users for future multiplication.

Basically, plants are produced by sexual and asexual or vegetative methods. In India, sexual method is mostly followed in flowering annuals (seasonal flowers). Besides these, many shade, flowering or ornamental trees, palms, cacti, succulents as well as some annual creepers (like *Clitoria ternatea*, *Lathyrus odoratus*, *Ipomea* spp. and *Thunbergia alata*) are also produced from seeds. Breeders often bred several ornamentals (like rose, chrysanthemum, gladiolus, carnation, orchid, dahlia, bougainvillea and hibiscus etc.) through seed only to get new hybrids. On the other hand, many ornamental plants are propagated through vegetative means (like cutting, layering, buddings), and also through corms, bulbs and tubers. The following paragraphs will describe vegetative propagation methods of some common flowers crops:

5.1 Chrysanthemum

Chrysanthemum is propagated vegetatively either through sucker, cutting or through micro propagation. After flowering, the stem is cut back first above the ground. This induces the formation of side sucker, which are separated from the mother plant and are planted in sand bed. Well rooted sucker can be used for planting. Cuttings are taken from a healthy stock plant. 5-7 cm long cuttings are made by shearing basal leaves and cutting half of the open leaves. The cuttings are dipped in 2500 ppm Indole butyric acid (IBA) or either in Seradix/ Rootadex A or No.1 / Karadex (rooting hormones). These cuttings are put in sand beds in semi shade conditions and watered immediately and thereafter in a regular manner.

Cutting operations are sometimes done in a greenhouse for better result. The temperature within a greenhouse should be between 15-18^{0}C and that of the rooting medium be between 18 and 21^{0}C. A square meter of medium should accommodate 500-600 cuttings, depending on size of the lower leaf of the cultivar. Fine misting is done intermittently on the cuttings during day light hours. The mist is usually turned off a day or two before cuttings are removed for hardening. Cuttings are well rooted in 10-20 days depending on cultivation and season. Cuttings with roots of about 1.5-2.0 cm long are desirable since longer roots makes planting difficult.

5.2 Carnation

A typical carnation cutting (having a length of about 10-15 cm with 4-5 pairs of leaves weighing about 10 g) should be planted at 5 cm spacing in rooting medium. The rooting medium consists of peat moss and perlite (1:2) alongwith sufficient calcium carbonate to bring the pH near 7.0. A rooting hormone should also be used. Full rooting appears in about 21 days at a rooting temperature of 15^0C. Maintainance of bottom heat at a constant 21^0C is found to reduce the rooting time to 15 days. Watering through intermittent mist on bright warm days at an interval of 10 seconds out of every 4-6 minutes is necessary.

5.3 Rose

Roses can be propagated by many methods *viz.*, cutting, layering, budding and grafting. Among all these, budding on a rootstock is widely practiced. T-budding is the most popular and commercially practiced budding. The most common rootstocks are *Rosa multiflora* (Karnataka, West Bengal and Bihar), *Rosa indica* (North India) and *Rosa bourbaniana* (North India), A thornless rootstock namely NISHKANT has been developed at IIHR Since it is devoid of thorns. Close planting of rootstocks can be taken up for budding operation.The rootstocks are normally propagated through cuttings. The dormant buds from a selected variety are carefully removed with a sharp knife along with a small portion of stem. The selected rootstock is pruned to about 7-10 cm height. On a pencil thick stem, a T-shaped cut is made and the bark is slightly loosened to accommodate the selected bud. After inserting the bud the bark portion of the rootstock is covered and tied with a film of polythene. The dormant buds get incorporated into the rootstock and sprout to produce new flush.

5.4 Orchid

Orchids may be propagated either sexually or asexually. Since, most orchids do not come true to type from seed, once a hybrid or clonal selection has been made, then all further propagation is done through asexual means for getting off-springs true to type.

Orchid seeds are very small. As the seeds are lacking with endosperm, they are difficult to germinate. Most monopodial orchids (*e.g.*, Vanda) can be propagated by tip cuttings. Vanda tip cuttings are usually 30-37 cm tall and bear up to 12 leaves and usually few aerial roots. Cuttings can be potted and will grow without being put in a propagation bed.

Some monopodial and sympodial orchids (such as Dendrobium and Epidendrum) produce off-sets in leaf axil. 3-4 root off-sets can be snapped off and potted up. Cattelya and other sympodial orchids are propagated by division of the parent clump. This is usually accomplished on plants that have six or more pseudo-bulbs. The rhizome is cut between the third and forth pseudo-bulbs and both sections are potted up as individual plants. Since, most cattleya plants produce only one new leaf per year, most plants are divided every three years. Paphiopedilum and Cymbidium can be divided more frequently, as a division containing only one fan of leaves or one pseudo-bulb is all that is necessary to increase these plants.

5.5 Gladiolus

Gladiolus corms are propagated from carmels which grow in clusters on outgrowth (stolons) between mother and daughter corms. Cormels are usually graded into three sizes (large 1.0 cm diameter; medium 0 6 cm and <1.0 cm; and small <0.6 cm). Mostly the large sizes of cormels are used for planting stock production. Carmel stocks should be chosen carefully to prevent the spread of disease into developing corms and preferably should only be from healthy, disease free and roughed block. The cormel should be treated in hot water (53-55^{0}C) for 30 minutes to eradicate fungus, insects and nematodes. Sometimes this treatment is combined with fungicides like benomyl (0.10 kg/100 litres water), captan (0.18 kg/100 litres) or thiram (0.18 kg/100 litres) to compliment the action of hot water. Two days prior to treatment, carmel should be covered with warm water (32^{0}C) to soften the husk. The treated cormels should be air-dried in thin layers in sterilized trays and then placed in cold (2-4^{0}C) until planted. Dormancy of large cormels is usually broken within four months of treatment. Root bud swellings indicate that cormels are ready to be planted. It is a good practice to soak cormels in water for 2 days just prior to planting to ensure uniform sprouting. Corms from 1.3 to 2.5 cm diameter are called "planting stock" and are used for the production of flowering size corms.

5.6 Tuberose

Tuberose is multiplied through bulbs (of size 1.5cm and above diameters). 8-10 bulbs per metre of rows are planted (at a spacing of 30 cm x 30cm). The bulbs are placed 5 to 8cm depth over a ridge or in a flat bed. The bulbs sprout 10-15 days after the planting, depending upon the temperature. Timely irrigation, weeding and broadcasting of nitrogenous fertilizer is practiced to maintain good growth. The bulbs are snapped off from the clump and kept in shade for two to three days. The soil is removed from the bulbs and they are subjected to storage in normal ventilated conditions. The multiplication of bulbs ranges from 1:10-15 times.

5.7 Bougainvilleas

Bougainvilleas are propagated by cuttings, ground or air layering and budding. Normally pencil thick hardwood leafy cuttings (15-25cm long) are preferred. They are treated with growth regulator such as IBA (Indole butyric acid @) 1,000 and 3,000 ppm in the quick –dip- method in solution) or talc. The cultivars, which do not respond to propagation by cuttings, are raised through layers. In air layering, the media used is mold or leaf mold with farmyard manure or farmyard manure with soil and sand. T or shield budding is followed for the varieties, which do not respond to above methods. Generally Dr. R.R. Pal (a robust cultivar) is used as rootstock. The best time for budding is February to March.

5.8 Jasmine

The commercial method of multiplying Jasmine is cutting. However, propagation by layering and even by grafting (approach or inarching) and budding is also possible. To facilitate the rooting in layering (ground or air layering) or cuttings, 1000-2000 PPM of IBA is applied to the basal portion for rooting. Fifteen centimeters long shoot tip cutting with four leaves and five distal buds are placed in a rooting media of vermiculate or good soiled and then rooted in a mist chamber. In the open, hardwood or semi-hardwood (15-20 cm long) cuttings with or without leaves are used for multiplication. The hardwood cuttings of *J. sambac* may be planted directly *in situ* during rainy season by which a success of 70-80 per cent rooting may be obtained. Layering of tender shoots ensures better and quick rooting and multiplication is done in rainy season or June-July. High humidity is maintained to promote rooting of the cutting.

5.9 Canna

It can be easily propagated by division of rhizomes. This rootstock is a branchy mass with many large buds. While dividing the rootstocks, it is better not to cut too close but to leave several strong buds on each piece and may be planted directly in the beds. If stock is not abundant, only a few plants should be made retaining at least one bud on each piece. It will be worthwhile to bear in mind that weak buds produce only weak plants. The one-bud plants are usually better suited for planting in pots. Propagation by seeds in quite a difficult process and is normally not adopted unless and until it is for the purpose of raising new hybrids.

5.10 Some Important Ornamental Plants: Modes of Propagation

Name	Propagation
Common garden plants and trees	
Bauhinia	Seeds
Celiba pentandra (Silk Cotton)	Seeds
Cassia excelsa	Seeds
Cassia javanica	Seeds
Erythrina indica	Cutting
Glyricidia maculata	Seeds
Jacaranda mimosifolia	Seeds
Lagertemia flosreginae	Seeds
Michelia champaka	Seeds and grafting
Gulmohar	Seeds
Peltophorum	Seeds
Foliage trees (ever green)	
Alstonia	Seeds
Araucaria (Monkey puzzle)	Seeds
Artocarpus (Bread fruit)	Seeds
Casuarinas	Seeds
Thuja	Seeds and cutting
Felicium decipens (fern leaved tree)	Seeds
Ficus benjamina	Seeds
Ficus elastica	Cuttings, Air layering
Grevilea robusta (Sliver Oak)	Seeds
Kigelia pinnata	Seeds
Polyalthia longifolia	Seeds
Royal palm (*Oreodoxa regia*)	Seeds

Contd

Table 5.10 Contd...

Name	Propagation
Flowering shrubs	
Achania	Cuttings
Barleria	Cuttings & Seed
Buddleacaesalpinia pulcherrima (Peacock flower)	Cuttings & layers
Cestrum nocturnum (Night Queen)	Seeds
Hibiscus	Cuttings
Ixora	Cuttings & layers
Jatropha	Cuttings & Seed
Lantana	Cuttings & layers
Lagerstroemoea indica	Cuttings & layers
Nerium oleander	Cuttings & layers
Poinsettia	Cuttings & layers
Tabernaemontana	Cuttings & layers
Rose	Cuttings & bidding
Thevetia	Cuttings & Seed
Foliage shrubs	
Acalypha	Cuttings
Anthurium	Rhizomes
Aralia	Cuttings
Arunda donax	Suckers
Coleus	Cutting
Crotons	Cuttings, layers
Dieffenbachia	Cuttings & Suckers
Dracaena	Cuttings
Duranta	Cuttings
Eranthemum	Cuttings
Maranta	Cuttings
Panax	Cuttings
Iresene	Cuttings
Phyllanthus (variegated)	Cuttings & Suckers
Creepers	
Allamanda	Cuttings
Antigonon	Seed, Tubers
Aristolochia elegans	Seeds & Layers
Bignonia gracillis	Cuttings & Layers

Contd

Table 5.10 Contd...

Name	Propagation
Bigonia venusta	Cuttings & Layers
Clematis	Cuttings
Clerodendron	Cuttings
Echites	Layering
Honeysuckle	Cuttings
Ipomea horsfolleae	Cuttings
Monstera edulis	Cuttings
Passion flower	Cuttings
Pothos	Cuttings
Quisqualis indica	Root suckers
Flowering annuals	
Aster, Snapdragons, Marigold, Celosia, Cosmos,	
Comphrena, Helianthus annum Helichrysum,	
Impatients, *Lobella verbena*, Plox, Zinnia, etc.	Seeds

Chapter 6

Production of Hybrid Seeds

In simple terms, hybrid seed is seed produced by cross-pollinated plants. It is done by crossing between two genetically dissimilar parents. Pollen from male parent (Pollen parent) will pollinate, fertilize and set seeds in female (seed parent) to produce F1 hybrid seeds. Hybrids are bred to improve the characteristics of the resulting plants, such as better yield, greater uniformity, improved colour, disease resistance, *etc.*

In nature, to create genetic variability and for its wider adaptation in different environmental conditions, flowering plants has adopted many mechanisms for cross pollination. Cross-pollination results in genetic heterogeneity and show wider adaptations. Flowering plants have evolved a number of devises to encourage cross-pollination. Those mechanisms are;

1. Dicliny: Flowers are unisexual. In monoecious plants male and female flowers are borne on the same plant *e.g.,* cucurbits, coconut,etc. In dioecious plants male flowers are borne on different plants *e.g.,* papaya, cannabis, *etc.*

2. Dichogamy: Time of anther dehiscence and stigma receptivity are different forcing them for cross-pollination. In protoandry anthers dehisce earlier than the stigma receptivity. In protogyny stigma become recetive earlier than the anther dehisce.

Examples: Plants of the families Cruciferae, Rosaceae and Ranunculaceae.

3. Self-incompatibility: Self fertilization in avoided by recognizing the self pollen by the stigma. *E.g.*, Petunia, Lilium .

4. Herkogamy: There is spatial separation of the anthers and stigma. Their relative position is such that self fertilization cannot occur. The stigma projects beyond the anthers and therefore pollen cannot land on stigma. *e.g.*, Lucerne stigma is covered with a waxy film. The stigma does not become receptive until this waxy membrane is broken by visit of honeybees resulting in cross-pollination.

5. Male sterility: Absence or atropy or mis or malformed of male sex organ (functional pollen) in normal bisexual flower. Male sterility is of three types: genetic male sterility, cytoplasm sterility and cytoplasmic-genetic male sterility.

6. A combination of two or more of the above mechanisms may occur in some species. This improves the efficiency of the system in promoting cross-pollination

6.1 Requisites of hybrid seed production:

1. **Breeders responsibilities:**
   (a) Develop inbred lines
   (b) Identification of specific parental lines
   (c) Develop system for pollen control

2. **Major problems for breeders and producers**
   (a) Maintenance of parental lines
   (b) Separation of male and female reproductive organs
   (c) Pollination

6.1.1. Basic procedures for hybrid seed production

1. Development and identification for parental lines
2. Multiplication of parental lines
3. Crossing between parental lines and production of F_1

Commercial hybrid seed production demands crossing technique which is easy and also economic to maintain parental lines. Only few crossing mechanisms have been adopted for commercial hybrid seed production they are;

1. Hand emasculation and pollination
2. Self-incompatibility

3. Dicliny : monoecious and dioecious

4. Male sterility

These techniques are specific to crop floral biology and flowering behaviour. These techniques have their own advantages and disadvantages. Based on the crop behaviour and crossing technique have been adapted for production of hybrid seeds commercially.

The production of F_1 hybrids in vegetables and flowers can be attractive to breeders but introduces special problems of seed production due to the breeding system and methodology. The initial breeding of hybrids is costly and the seed production costs are also comparatively high. Furhermore, with our present knowledge of F_1 seed production, it is not possible to prevent the loss of entire seed crops on certain occasions owing to environment factors, or to avoid the need to destroy a crop because of excessive sib proportions. The cost of these failures has to be borne by the successful crops (Willis and North,1978).

6.2 F_1 hybrid seed production in vegetable crops

The trend of F_1 hybrid seed usage in vegetable crops is increasing globally in term of species, cultivars and volume of seed used. Most of the seed of our main vegetables including tomato, sweet pepper, eggplant, cucumber, squash, pumpkin, melon, watermelon, brassicas such as cabbage, cauliflower, broccoli, Chinese cabbage and radish, and onion in developed countries are of F_1 hybrid cultivars. The popularity of F_1 hybrid cultivars is due to their vigour, uniformity, disease resistance, stress tolerance and good horticultural traits including earliness and long shelf-life expressed and therefore giving consistent stable high yield. From the breeder point of view, it is a fast and convenient way to combine desirable characters of a vegetable together, for example fruit size and colour, plant type and disease resistance, and as a mean to control intellectual property rights through control and protection of the parental lines by the breeders.

Vegetables can be classified into three categories according to their temperature requirements as follows:

- Low temperature species such as brassicas, radish, carrot and spinach require a low temperature of 8-15°C of vernalization to bolt, flower and seed set

- Moderate temperature species such as tomato, sweet pepper and zucchini require a temperature of around 18-20°C with around

25°C in the day and 15°C at night for optimum seed production. The diurnal temperature difference is desired to obtain best result. Too low temperature causes low seed set and pollen production, and too high temperature causes flower abscission, low pollen production and viability, pest and disease problem; and

- High temperature species such as okra, cucurbits, sweet corn and tropical vegetables require a temperature of 20°C and above.

In the tropics seed production is sometimes achieved in highlands of 500-2000m altitude where the cool temperature is suitable for the moderate temperature loving vegetables to produce seed. Photoperiodic reaction is not a concern as most of the modern cultivars of the moderate temperature requirement species are day-neutral plant and thus insensitive to photoperiod. The low temperature requirement vegetables often require a specific period of cold vernalization in their growth phase to induce bolting and flowering. Seeds of these species are often produced in the higher latitudes.

6.3 Hybrid varieties and saving seed

Most horticulturists and plant breeders often discourage home gardeners to grow their own seed. This is mainly to ensure the best quality seed for their intended crop with satisfactory performance. The following are some of the precautions to be taken care of particularly for hybrid vegetable seeds.

1. Seed from hybrid vegetables should not be saved because they won't produce true in the next generation. An F_1 hybrid is the result of crossing two pure lines to achieve the desired result. Scientific and accurate breeding programs have made it possible not only to bring out the outstanding qualities of the parent plants, but in most cases, these qualities have been enhanced and new desirable characteristics added to the resultant hybrid plants. In addition to qualities like good vigour, true- to-type, heavy yields and high uniformity which hybrid plants enjoy, other characteristics such as earliness, disease resistance and good holding ability have been incorporated into most F_1 hybrids. Uniform plant habit and maturity, coupled with uniformity in shape or size have made hybrid vegetables extremely suitable for mechanical harvesting.

2. It is difficult for the common grower to isolate varieties and strains to avoid unwanted cross-pollination. Cross-pollination can be a major problem if the grower works in the midst of many other gardens where he has no control over what is being grown around him.

3. Unwanted cross-pollination and faulty selection of parent plants result in the gradual deterioration or "running out" of the seed. In the case of seed saving, a part of the row or maybe a few plants in the row are tagged as those to be allowed to produce seed. The vegetables of designated plants will be allowed to remain until mature on the plant.

Following are some simple directions on how to save seed from some of the most commonly grown garden vegetables.

Beans (all kinds): The seed is allowed to mature thoroughly on the plant, usually indicated by size of the seed in the pod or by the colour of the pod. The entire plant is to be pulled early in the morning and placed it in the shade to dry out. This will prevent the pods from split open and the beans from shattering.

Cucumbers: Cross pollination in cucumbers helps the pollen to be transferred from a plant of one variety to a plant of another variety. This is done by insects. The plants that come from these seeds as well as the fruit will be different. So, only one variety is to be planted selected from strong, healthy cucumber plants with well-formed fruits.

Eggplant: When the eggplant fruit has obtained maximum size and shows some evidence of browning and shriveling, it is ready to be harvested for seed. The seeds are to be removed and washed thoroughly to remove all pulp, then to be spread out in the sun to dry quickly (as moist seed will begin to germinate overnight if left in a damp condition). Storage in a cool, dry place is advocated.

Okra: Okra pods should be left on the stalk until brown and well matured. The pods are removed and placed in the shade until thoroughly dried. Although the seed may be removed from the pod, it is generally best to store them in the pod until ready for planting at which time the pods may be split open and the seed removed. Pods harvested too green will not store well and are likely to split, shattering the seed.

Peppers: Pepper should be allowed to ripen until they become red. The pepper pods are cut into halves and the seed are scraped from a cavity onto a piece of paper. The seeds are spread out and dried thoroughly before placing in a storage container.

Southern Peas: Southern peas should be left on the plant until thoroughly matured, usually indicated by a browning of the pods. The pods should be picked, spread out in a dry area and cured for a week or two, then shelled.

Squash: If seeds are to be saved from squash, only one variety is grown in the garden. When the outer covering of the squash has become hardened, the seed are generally mature. The squash fruits are split open, then the seeds are scoop out and washed until all pulp is removed. To dry the seeds they are spread out on newspaper.

Tomatoes: The tomato fruits are allowed to ripen thoroughly on the vine. The fruits are cut open and the seeds are removed by squeezing or spooning out the pulp with seeds into a non-metal container such as a drinking glass or jar. The container is set aside for one or two days. The pulp and seed covering will ferment so that the seeds can be washed clean with a directed spray of water into the fermented solution. The clean, viable seeds will drop to the bottom of the solution, allowing the sediment to poured off. Several rinsings may be necessary. Then the tomato seeds are spread out on a cloth or paper towel to dry. After seed are dry, they are packed, labelled with date for storage in a cool (refrigerator), dry location.

Chapter 7

Future Thrust, Recommendation and Conclusion

Seeds and planting materials are the most important inputs for the success of horticulture development. For ensuring availability, quality and health of seeds and planting material in horticultural crops about 20 million hectares of agricultural land have already covered under horticulture crops, which result in yield of 91 million tonnes. Though these crops occupy only around 10% of the cropped area, they contribute over 18% to the gross agricultural output in the country. Moreover India has developed more than 1,500 cultivars of horticultural crops of which, many are adopted by farmers. Most of the horticulture crops are vegetatively propagated and there is risk of transmission of diseases. To meet the increasing demands of quality seeds and planting material we need to take care of some basic things. Diagnostic techniques are to be used in a larger scale and phyto-sanitary standards are to be followed strictly for early detection of diseases in many fruit and vegetable crops (*viz.*, banana, citrus fruit, potato and other tuber crops) to weed out unhealthy propagating material. There is a need to work out on certification standards for different horticulture planting materials. Quality standards should be specified and certification should be enforced. With the entry of multinational companies in the seeds sector, especially in research and development of Genetically modified (GM) seeds and crops, bringing quality assurance of cloned seeding material of GM horticultural crops under legal regulatory regime has become imperative. There is need for legislative framework and enforcement for quality assurance of vegetatively propagated plants. With the changing scenario of global warming, the cultivars suitable for wider adaptation like high temperature, drought and salt tolerance needs to be developed.

7.1 Current status of Seeds and Planting Materials– National and International Scenario

The following policy initiatives have been taken by the Government of India in seed sector:

- Enactment of the Seeds Act, 1966
- Seed Review Team-SRT (1968)
- National Commission on Agriculture's Seed Group (1972)
- Launching of the World Bank aided National Seeds Programme (1975-85) in three phases leading to the creation of State Seeds Corporations, State Seed Certification Agencies,State Seed Testing Laboratories, Breeder Seed Programmes etc.
- Seed Control Order (1983)
- Creation of the Technology Mission on Oilseeds and Pulses (TMOP) in 1986 now called The Integrated Scheme of Oilseeds, Pulses, Oil Palm and Maize (ISOPOM).
- Production and Distribution Subsidy
- Distribution of Seed Mini-kits
- Seed Transport Subsidy Scheme (1987)
- New Policy on Seed Development (1988)
- Seed Bank Scheme (2000)
- National Seeds Policy (2002)
- The Seeds Bill (2004)
- Formulation of National Seed Plan (2005)
- National Food Security Mission (2007)
- Rashtriya Krishi Vikas Yojna (2007).

Seed and propagating material - EU rules

The EU regulates the marketing of seed and propagating material of agricultural, vegetable, forest, fruit and ornamental species and vines, ensuring that EU criteria for health and quality are met. EU legislation applies to genera and species important for the internal market and is based on:

- Registration of varieties or material;
- Certification or inspection of lots of seed and plant propagating material before marketing.

Seed from non-EU countries has to meet equivalence criteria. Non-EU countries guaranteeing they meet these criteria are listed as eligible to export to the EU.

Under certain conditions, EU countries may temporarily market seed which does not meet these standards *e.g.* when a supply shortage occurs mostly caused by reduced germination of seed. The Commission establishes the rules in collaboration with national experts in the respective Standing Committees.

Vegetable seeds

In the EU, seed of registered vegetable varieties can be marketed once officially examined and certified. In some cases, seed of a category not officially certified can be checked as meeting the legal requirements. Before marketing in the EU, vegetable varieties must be listed in the Common catalogue based on EU countries' national catalogues. Only distinct, stable and sufficiently uniform varieties are accepted. Suppliers must ensure that their material meets the legal criteria.

Fruit propagating material and fruit plants

Fruit plant propagating material and fruit plants of 23 genera and species may only be marketed if certified as: Pre-basic; Basic; Certified and Qualifying for *Conformitas Agraria Comunitatis*. To gain this certification, official inspections check if the material meets criteria for: Identity; Quality and Plant health. The rules also cover batch separation and marking, identification of varieties and labelling. Suppliers must ensure that their material meets the legal criteria. Only suppliers approved by their national authorities can sell their material in the EU. Seed, propagating and planting material from non-EU countries may only be marketed in the EU if they offer the same guarantees as those produced in the EU.

Ornamental seeds and plants

The seed and propagating material of ornamental plants can only be marketed if it:

- Substantially free from harmful organisms that may affect its quality as propagating material;
- For propagating material - has satisfactory vigour and dimensions;
- For seed - has satisfactory germination.

Suppliers are responsible for the quality of their products. Before marketed in the EU, the material has to meet legal criteria such as:

- Batch separation and marking; Accurate identification of varieties if needed and Labelling.

7.2 Suggested Research Directions

Technology generation for production of vegetatively propagated plants: Plant growth regulators (PGRs) have significant role in vegetative propagation of certain plants. Technologies involving such PGRs may be improved further in respect of efficiency and new technologies to be developed for propagating difficult-to-root and incalcitrant horticultural plants species by exploring synergism among different PGRs and also between them and certain phenolics. Uniformity in certification programmes is another area to be taken care of. Immediate attention is required for the dissemination of innovative technologies to nursery men and their capacity building.It is important to produce quality planting material of ornamental crops through clonal multiplication and *in vitro* micro-propagation. Sensitive diagnostic assays such as ELISA and PCR for indexing and certification of clonally propagated and tissue culture plants have to be used. A simple PCR-based diagnostic system suitable for detection of leaf spot causing pathogens such as *Alternaria*, *Colletotrichum* and *Phytophthora* may be used for latent infection detection in planting material. Healthy and quality planting material should be supplied by public institutions (Institute/SAUs) in large quantities as nucleus planting material.

Production of seeds of vegetables and seed spices crops: There is need to assess all available variability in the form of genetic resources/ land races in different agro-climatic zones for their possible exploitation in crop improvement programme. The problems of quality seed production should be overcome through advancement made in technology of hybrid seed production, use of plastic culture etc. To ensure quality seeds, balanced use of several macro and micronutrients alongwith other nutrients, organic manures, biofertilizers, bioagents and plant growth regulators are suggested. The production of quality seed of major vegetable crops and seed spices should be given priority in collaboration of ICAR Institution, SAU's and private sector of vegetable seed industry. The certification of quality seed based on quality parameter should also be needed to advocate to progressive vegetable growers and farmers engaged in production of vegetable crops and seed spices towards quality seed production.

Production of tissue culture plants and their management: Research initiatives are to be taken for the development of reproducible and efficient protocols for recalcitrant plant species such as cashew, coconut, etc. alongwith low cost and sensitive diagnostics for various diseases including viral diseases. Use of bar code system and software for ensuring the genetic purity of tissue culture plant should be introduced in commercial micropropagation.Public funded institutions have already developed several laboratory scale protocols. There is a need for public-private partnership to commercialize these protocols. State-of-the-art commercial level micropropagation facilities should be developed in horticultural institutions (one each in north, south, west and east).

Production of Quality Planting Material of Ornamental Crops: Tissue culture procedures should also be employed in all ornamental crops, wherever feasible. Protocols already available for micropropagation of flower plants may be refined and tried on mass scale. Diagnostics kits for detection of viruses for important ornamentals should be available. Bulb production of important bulbus ornamental should be standardized in order to reduce their import. Works on developing protocols for difficult to propagate flower plants and also for new and exotic materials are to be initiated. State-of-the-art infrastructure facilities like tissue culture lab, polyhouses, mist chambers, net-houses, microirrigation, fertigation, hardening facilities etc. are to be created on urgent basis. The nursery propagation techniques for raising the mother plants and the secondary materials are to be refined. Appropriate media and containers for micro-propagated flower and foliage plants are to be standardized.

Production of Quality Planting Material of Plantation and Spices crops: The technologies for quality planting material production of plantation and spices crop need to be up-scaled through 'Seed Village Concept' with the help of KVK for augmenting the production. In coconut, 'plumule culture' technique needs to be refined for up-scaling the production of homogenous parental lines of coconut for production of quality planting material. Production of seedlings of specific varieties of coconut needs to be strengthened for establishing gardens for promotion of product diversification. Enriching the potting mixture with bio-fertilizers and bio-priming of seedlings with bioagents are to be recommended for healthy planting materials production. Technologies for micro-rhizome production have to be adopted to augment the diseases free 'seed rhizome' production in rhizomatous crops like ginger and turmeric. Replanting programme is essential to replace the senile and old plantations with quality and healthy materials in these crops. Establishment of mother gardens and scion block with high density planting is highlighted in cashew, black pepper and tree spices and has to be practiced for increasing the production of planting material. Different diagnostic methods of diseases and pests of these crops are available and these 'diagnostic kits' are to be popularized to detect the problems at field/nursery level. Location specific varieties have to be planted for meeting regional scale demands. Micro-propagation is to be strengthened as an important option to generate quality nucleus planting materials. Accredited Labs are available to test viruses and genetic fidelity with biotechnological tools that has to be made use of for enhancing quality in the system of production.

Regulations, Policy Initiatives and Partnership for Enhanced Availability of Seed and Planting Material: There should be no compromise on issues of efficiency and accountability, be it the public sector or private sector. Public and private partnership in-R&D horticulture is happening but it needs to be augmented. Functional efficiency and accountability in public sector developmental organizations needs to be further improved and to harness synergies, an appreciation of the new IP regime is required in both sectors. There is a need to establish state of the art Quarantine Laboratories for inspection of all seed and planting material exchanged indigenously within the country, as well as to intercept pest and diseases in material received from exotic sources. There is a need to upgrade the quality of farmers' saved seeds through hi-tech interventions, besides using such interventions in the production of quality seed and planting material in horticultural crops. A need is also felt to encourage Public Private Partnership (PPP) in establishment of greenhouse technology for production of healthy planting material.

Outscaling innovation and adoption of cutting edge technologies, such as biotechnology and nanotechnology, would be critical for desired impact on livelihood of resource poor small holder farmers. The Indian seed sector today is well established with tremendous potential to grow beyond boundaries of domestic market. We must be proactive to explore export potential and create enabling environment (Paroda, 2013).

Bibliography

Agarwal,R.L.(1980)*Seed Technology*. Oxford and IBH Publishing Co. New Delhi.

Aggie Horticulture (2013).http://aggie horticulture.tamu.edu/archives/parsons/vegetables/seed.html Accessed on 25.08.2013

Agriwatch(2013).http://www.agriwatch.com/spices/dhaniya-coriander-seed/. Accessed on 29.08.2013

Anwar,F.; Rashid,U.;Ashraf,M. and Nadeem,M.(2010). Okra (*Hibiscus esculentus*) seed oil for biodiesel production. Applied Energy, 87(3): 779–785.

Arya (1983) Indian Farming Digest. 16: 13-16.

Bailey, L. H. 1949. Manual of cultivated plants. McMillan Co. NY.

Banga, O., 1976: Radish: Raphanus sativus Cruciferae. Evolution of Crop Plants N W Simmonds ed: 0-62

Bhonde, S.R.; Lechhiman, R.; Srivastava, K.J. and Pandey, U.B.(1989) A note on effect of spacing and levels of nitrogen on seed yield of onion. *Seeds Farms*,15(1):21.

Bienz D.R 1980. *The why and how of home horticulture*. W.H. Freeman and Company, New York.

Brewster, J.L. (2008) *Crop Production Science in Horticulture, Volume 15: Onions and other ve.getable* Alliums. Wallingford, Oxon, Great Britain: CABI Publishing. 2nd ed.

Chauhan, DVS (1981)*Ve.getable Production in India,3rd ed.* Ram Prasad and Sons, Agra and Bhopal, India.

CMIE, 2002 http://www.cmie.com/ Centre for Monitoring Indian Economy Pvt. Ltd.

CTCRI (1983). http://www.ctcri.org/publications.html

DGCIS, 2013. Directorate General of Commercial Intelligence and Statistics, Kolkata; http://www.dgciskol.nic.in/data_information.asp

Dravid,P.S. (2011). Future growth drivers for Indian Seed industry.*Indian Seed and Planting Material*. 4(4): 41-44.

Heiser, C. B. and P. G. Smith. 1953. The cultivated *Capsicum* peppers. Econ. Bot., **7**: 214-27.

http://indiaagrifarms.com/2011/12/chilli_cultivation/ Accessed on 29.08.2013

http://www.agrocrops.com/red-dry-chillies.php Accessed on 29.08.2013

Jayabarathi,M.;Palaniswamy,V.;Kalavathi,D.andBalamurugan,P.(1990).Influence of harvesting conditions on the yield and quality of brinjal seeds.*Ve.g. Sci.*,17(2):113.

Khah, E.M. and H.C. Passam. 1992. Flowering, fruit set and development of the fruit and seed of sweet pepper (*Capsicum annuum L.*) cultivated under conditions of high ambient temperature. *Journal of Horticultural Science* 67(2)251-258.

Kissan Kerala (2013)http://www.kissankerala.net/kissan/kissancontents /coconut.htm. Accessed on 25.08.2013

Macdonald,M.B. (1980) Assessment of seed quality. *Hort Science*, 15: 784-788.

Moekchantuk T, Kumar P (2004): Export okra production in Thailand. Inter-country programme for vegetable IPM in South and SE Asia phase II Food and Agriculture Organization of the United Nations, Bangkok, Thailand

NCPAH(2013) National Committee on Plasticulture application in Horticulture, http://www.ncpahindia.com/tomato.php. Accessed on 26.08.2013

OSU (2013). Department of Horticulture and Crop Science,The Ohio State University. http://www.hcs.osu.edu

Panigrahi,U.C.; Pattanayak,N.B. and Das,C. (1990)A note on the effect of micronutrients on yield of cauliflower seeds in the acid red soil of Orissa. *Orissa J. Hort,* 18(1-2): 62.

Paroda, R.S, (2013) Indian seed sector: The way forward. Special Lecture delivered at Indian Seed Congress 2013, Friday 8th Feb, 2013, Gurgaon,

Haryana; Organised by National Seed Association of India (NSAI), New Delhi

Popovska,H.P.;Miadenovski, L.T. and Mihajlovski,M.(1981)The influence of packing over germination of pepper and tomato seeds. *Acta Hort.*911:281.

Raj,D. and Kanwar,B.B. (1990) Minimizing insecticide use against cauliflower pests in India.*Trop.Pest Manage*, 36(1):10.

Rawat,T.S. and Singh,V(1981)Effects of spacing and nitrogen application on the performance of seed crops of radish cv. Pusa Rashmi. *Udyanica*, 4:17.

Salunkhe, D. K., Desai, B. B. and Bhat, N. R. (1987). Vegetable and Flower Seed Production. Agricole Publishing Academy. New Dheli.

Sambandamurthi and Sundaram (1989) A note on seed production of cabbage under Kodaikanal conditions. South Ind. Hort., 137(3): 183-186

Selvaraj,J.A. and Ramaswamy,K.R.(1988) Effect of density grading on seed quality attributes in brinjal. *Seed Res.*16(1):117.

Selvaraj,J.A.(1988)Studies on storage of brinjal seeds.I.Biocide treatments and containers for storage. *South Indian Hort.*36(6):313.

Singh SP. 2001. Seed Production of Commercial Vegetables. Agrotech Publ. Academy.

Singh, H. B., Bhagchandani, P. M. and Thakur, M. R. (1964). Indian J. Hort., 21 : 221-231.

Tate, K.G. and Cheah, L.H. (1983) Control of clubrot in cauliflower. *N.Z. Commer. Grower*, 38(9):36.

Vanagamudi, K.; Subramanian, K.S. and Baskaran, M. (1990) Influence of irrigation and nitrogen on the yield and quality of chilli fruit and seed. *Seed Res.*, 18(2):114.

Wills, A.B. and North, C. 1978. Problems of hybrid seed production. Acta Hort.(ISHS) 83:31-36. http://www.actahort.org/books/83/83_3.htm

http://agriexchange.apeda.gov.in/indexp/monthexport.aspx

National Seed Association of India NSAI (2005) http://nsai.co.in/

http://www.gmwatch.org/gm-firms/10558-the-worlds-top-ten-seed-companies-who-owns-nature

National Horticulture Board (2013). http://nhb.gov.in

https://www.facebook.com/notes/saving-us-all/a-list-of-gmo-and-non-gmo-seed companies/334086699976625

SeedNet India(2013).www.seednet.gov.in/Material/Seed_ Companies_ Pvt_FS_CS.pdf

Agriscape CompaniesSeeds(2013).http://www.agriscape.com/ companies/seeds/

Vegetable Hybrid Seed Production - CISST Seed Consortium

http://www.seedconsortium.org/PUC/pdf%20files/ 23Vegetable%20Hybrid%20Seed%20Production.pdf

U.S. Government Printing Office Home Page www.gpo.gov (Accessed on 13.09.2013)

Seed Testing and Accreditation Schemes.http://www.amseed.org/ resources/seed-testing-accreditation-schemes/

Seedbank http://seedbank.dacnet.nic.in/Reports/crvSSCReport.aspx

Plant quarantine India. http://www.plantquarantineindia.org/

ICRISAT (2013). http://www.icrisat.org

Department of Biotechnology (2013).http://dbtbiosafety.nic.in

Government of India, Department of Agriculture and Cooperation .http:// /agricoop.nic.in

Department of Horticulture and Crop Science-The Ohio State University http:/www.hcs.osu.edu

http:/www.hcs.osu.edu

Appendices

Appendix-1

Export Statistics : Export summary for Financial Year 2013-14
(Apr - April)

Product	Quantity(mt)	Value (Rs. lacs)
Fruit and Vegetable Seeds	1,434.40	2,680.68
Floriculture	1,177.82	2,382.15
Total	2,612.22	5,062.83
Fresh Grapes	39,653.30	36,532.59
Fresh Onions	1,32,250.95	20,628.56
Fresh Mangoes	9,634.36	7,012.61
Other Fresh Fruits	15,006.74	6,253.78
Other Fresh Vegetables	17,766.98	3,976.28
Walnuts	593.69	2,345.99
Total	2,14,906.02	76,749.81
Other Processed Fruits and Vegetables	31,986.47	17,469.23
Pulses	25,863.25	14,102.31
Dried and Preserved Vegetables	8,547.31	6,919.04
Mango Pulp	7,947.04	3,997.90
Total	74,344.07	42,488.48
Buffalo Meat	85,190.31	1,40,946.59
Dairy Products	12,068.31	20,850.90
Natural Honey	4,727.68	6,700.28

Product	Quantity(mt)	Value (Rs. lacs)
Sheep / Goat Meat	1,973.44	5,799.69
Poultry Products	791.71	1,629.48
Animal Casings	43.45	293.08
Processed Meat	34.75	76.32
Total	1,04,829.65	1,76,296.34
Guargum	51,182.67	1,56,970.49
Cereal Preparations	23,187.86	19,358.77
Alcoholic Beverages	29,487.35	18,103.25
Miscellaneous Preparations	27,387.62	16,318.03
Jaggery and Confectionery	3,951.29	2,781.08
Cocoa Products	705.19	2,014.45
Total	1,35,901.98	2,15,546.07
Basmati Rice	3,92,798.35	2,59,012.91
Non Basmati Rice	3,92,926.05	90,706.29
Other Cereals	4,91,692.88	76,817.31
Milled Products	23,468.82	5,199.06
Total	13,00,886.10	4,31,735.57
Grand Total	1833480.04	947879.1

Source: DGCIS Annual Export http://agriexchange.apeda.gov.in/indexp/monthexport.aspx

Appendix-2

A. New vegetable hybrids in India

Crop	Public Sector	Private Sector
Tomato	3	160
Eggplant	8	218
Chilli	2	73
Capsicum	1	31
Cauliflower	1	35
Cabbage	-	20
Okra	2	32
Watermelon	2	25
Cucumber	2	10
Gourds	6	80

Source: NSAI (2005)

B. The World's Top 10 Seed Companies

Company – 2007	Seed sales (US$ millions)	% of global proprietary seed market
Monsanto (US)	$4,964m	23%
DuPont (US)	$3,300m	15%
Syngenta (Switzerland)	$2,018m	9%
Groupe Limagrain (France)	$1,226m	6%
Land O' Lakes (US)	$917m	4%
KWS AG (Germany)	$702m	3%
Bayer Crop Science (Germany)	$524m	2%
Sakata (Japan)	$396m	<2%
DLF-Trifolium (Denmark)	$391m	<2%
Takii (Japan)	$347m	<2%
Top 10 Total	$14,785m	67% [of global proprietary seed market]

Source: ETC. Group [http://www.gmwatch.org/gm-firms/10558-the-worlds-top-ten-seed-companies-who-owns-nature]

Appendix-3

Area and Production of Horticulture Crops - All India

Crops	2010-11		2011-12	
	Area	Production	Area	Production
FRUITS				
Almond	23	14	22	4
Aonla	67	677	95	961
Apple	289	2891	322	2203
Banana	830	29780	797	28455
Ber	22	188	34	252
Citrus				
(i) Lime/Lemon	219	2108	234	2272
(ii) Mandarin	324	3255	329	3128
(iii) Sweet Orange (Mosambi)	157	1316	162	1232
(iv) Others	147	784	190	1290
Citrus Total (i to iv)	846	7464	915	7922
Custardapple	15	105	19	127
Grapes	111	1235	116	2221
Guava	205	2462	220	2510
Jackfruit	36	540	60	1042
Kiwi	0	1	3	6
Litchi	78	497	80	538
Mango	2297	15188	2378	16196
Papaya	106	4196	117	4457
Passion Fruit	–	–	16	97
Peach	18	92	20	91
Pear	41	300	48	294
Pineapple	89	1415	102	1500
Plum	14	32	26	72
Pomegranate	107	743	112	772
Sapota	160	1424	163	1426
Walnut	114	187	150	284
Others	913	5447	889	4991
Total Fruits	**6382**	**74878**	**6705**	**76424**

Contd…

Crops	2010-11		2011-12	
	Area	*Production*	*Area*	*Production*
VEGETABLES				
Beans	100	888	118	1151
Bittergourd	68	749	77	866
Bottlegourd	75	1354	105	1984
Brinjal	680	11896	692	12634
Cabbage	369	7949	390	8412
Capsicum	6	65	10	127
Carrot	56	953	62	1153
Cauliflower	369	6745	391	7349
Cucumber	35	525	40	607
Muskmelon	40	740	38	791
Okra	498	5784	518	6259
Onion	1064	15118	1087	17511
Peas	370	3517	408	3745
Potato	1863	42339	1907	41483
Radish	133	1878	160	2286
Sitaphal/Pumpkin	5	143	11	278
Sweet Potato	113	1047	110	1073
Tapioca	221	8076	227	8747
Tomato	865	16826	907	18653
Watermelon	67	1436	71	1727
Others	1496	18526	1661	19487
Total	8495	146554	8989	156325
Aromatic	510	605	506	566
Flowers cut*	–	69027	–	75066
Flowers loose	191	1031	254	1652
PLANTATION CROPS				
Arecanut	400	478	464	681
Cashewnut	953	675	979	725
Cocoa	57	14	63	13
Coconut	1896	10840	2071	14940
Total	**3306**	**12007**	**3577**	**16359**

Contd...

Crops	2010-11		2011-12	
	Area	Production	Area	Production
SPICES				
Ajwan	26	22	35	27
Cardamom	87	16	89	16
Chillies (Dried)	792	1223	805	1276
Cinnamon/*Tejpata*	3	5	3	5
Celery,Dill & Poppy	38	40	33	33
Clove	2	1	2	1
Coriander	530	482	558	533
Cumin	508	314	594	394
Fenugreek	81	118	94	116
Fennel	62	105	100	143
Garlic	201	1058	242	1228
Ginger	149	702	155	756
Nutmeg	16	11	17	13
Pepper	184	52	200	41
Vanilla	7	1	7	1
Tamarind	60	206	58	203
Turmeric	195	993	219	1167
SPICE Total	**2940**	**5350**	**3212**	**5951**
TOTAL	**21824**	**240427**	**23242**	**257277**

Source: *Horticulture Crop area statistics in India, by National Horticulture Board** Cut flowers in Lakh Nos.; Area and Production in '000 Ha and '000 MT, respectively.
Figure of Production under Grand Total does not include Production of Cut-Flowers.

Appendix-4

Approximate longevity (years) of selected vegetable and flower seeds.

Longevity (years)	Vegetable seeds	Flower seeds
5	Cucumber, Garden cress, Muskmelon, Radish, Water cress	Carnation, Chrysanthemum, Hollyhock, Nasturtium, Sweet sultans, Zinnia
4	Beets, Brussels sprouts, Cabbage, Cauliflower, Chicory, Brinjal, Fennel, Kale, Mustard, Pumpkin, Squash, Swiss chard, Tomato, Turnip, Watermelon	Sweet alyssum
3	Asparagus, Beans, Brussels sprouts, Carrot, Celery, Chervil, Chinese cabbage, Kohlrabi, New Zealand Spinach, Pea, Spinach	African daisy, Cosmos, Dusty miller, Marigold, Pansy, Petunia, Daisy, Snapdragon, Verbena
2	Dandelion, Leek, Sweet corn, Okra, Pepper	Aster, Phlox, Sweet pea
1	Lettuce, Onion, Parsley, Parsnip, Sea kale	Delphinium

Source: Adapted and modified from Bienz (1980).

Export of Horticultural produce in India

Products	2007-08 Qty (In MTS)	2007-08 Value(In ₹ Lakh)	2008-09 Qty (In MTS)	2008-09 Value (In ₹ Lakh)	2009-10 Qty (In MTS)	2009-10 Value (In ₹ Lakh)	2010-11 Value(In ₹ Lakh)	2010-11 Qty (In MTS)
Floriculture	36240.71	34014.42	30798.34	36881.41	26814.52	29446.36	28,645.41	28,645.41
Fruit and Vegetables Seeds	10157.13	14212.29	8535.53	11999.09	14507.51	14507.51	17,519.52	17,519.52
Floriculture & Seed Total	**46397.84**	**48226.71**	**39333.87**	**48880.50**	**43953.87**	**43953.87**	**46,164.93**	**46,164.93**
Fresh Onions	1008606.48	103577.89	1670186.29	182752.21	231942.98	231942.98	174,155.41	174,155.41
Other Fresh Vegetables	350235.47	48949.01	505285.46	68020.32	73185.9	73185.9	89,293.61	89,293.61
Dried Nuts (Walnuts)	6716.48	16207.80	5696.34	14123.63	19789.51	19789.51	15,650.59	15,650.59
Fresh Mangoes	54350.80	12741.76	83703.18	17071.25	20053.98	20053.98	16,292.13	16,292.13
Fresh Grapes	96963.57	31782.51	124627.97	40861.28	54533.89	54533.89	41,206.32	41,206.32
Other Fresh Fruits	207700.78	30452.60	256768.53	43086.84	52283.32	52283.32	48,964.74	48,964.74
Fresh Fruits & Vegetables Total	**1724573.6**	**243711.6**	**2646267.8**	**365915.53**	**451789.58**	**451789.58**	**385,562.80**	**385,562.80**
Dried and Preserved Vegetables	125726.28	42993.81	147861.22	49641.51	53207.48	53207.48	51,697.09	51,697.09
Mango Pulp	166752.17	50968.51	173013.60	75298.90	74460.77	74460.77	81,400.66	81,400.66
Other Processed Fruits and Vegetables	311756.29	96281.65	387126.42	137179.00	143550.63	143550.63	131,635.53	131,635.53
Processed Fruits and Vegetables Total	**604234.74**	**190244.0**	**708001.24**	**262119.41**	**271218.88**	**271218.88**	**264,733.28**	**264,733.28**
Grand Total	**2375206.2**	**482182.3**	**3393602.9**	**676915.4**	**766962.3**	**766962.3**	**696461.01**	**696461.01**

A List of GMO and Non-GMO Seed Companies

[https://www.facebook.com/notes/saving-us-all/a-list-of-gmo-and-non-gmo-seed-companies/334086699976625]

GMO Seed Companies

- American Seeds
- Asgrow
- Audubon Workshop
- Breck's Bulbs
- Burpee
- Campbell
- Cook's Garden -% of Monsanto Seed
- De Ruiter
- Dege Garden Center
- DeKalb
- Diener Seeds
- E & R Seed Co.
- Earl May Seed
- Ferry Morse – Possibly
- Fielder's Choice
- Flower of the Month Club
- Fontanelle
- Garden Trends
- Gardens Alive

- Germania Seed Co.
- Gold Country Seed
- Hawkeye
- Heartland
- Heritage Seeds
- Holdens
- HPS
- HPS Seed
- Hubner Seed
- Icorn
- J.W. Jung Seed Co.
- Johnnys Selected Seeds - 4% Monsanto seed
- Jung Seed
- Jungs
- Kruger Seeds
- Lewis Hybrids
- Lindenberg Seeds
- McClure and Zimmerman Quality Bulb Brokers
- Mountain Valley Seed
- Osborne
- Otis S.Twilley Seed Co.
- Park Bulbs
- Park Seed -% of Monsanto Seed
- Park's Countryside Garden
- Peotec
- Poloni
- R.H. Shumway
- Rea Hybrids
- Rocky Mountain Seed Co.
- Roots and Rhizomes
- Rupp

- Seeds for the World
- Seeds of the World
- Seminis Companies
- Seymour's Selected SeedsSnow
- Specialty
- Spring Hill Nurseries
- Stewart
- Stokes
- Stone Seed
- T & T Seeds, Ltd.
- The Vermont Bean Seed Company
- Tomato Growers Supply
- Totally Tomatoes
- Trelay
- Wayside Gardens
- Western Seeds
- Willhite Seed Co.

Non-GMO Seed Companies

- Abundant Life Seeds
- Amishland Seeds
- Annie's Heirloom Seeds
- Baker Creek Heirloom Seeds - rareseeds.com
- Baker Creek Seed Co.
- Berlin Seeds
- Botanical Interests
- Bountiful Gardens
- Cook's Garden -% of Monsanto Seed
- Diane's Flower Seeds
- Dirt Works
- Fedco Seed Co.
- Garden City Seeds
- Grannys Heirloom Seeds

- Harris Seeds
- Heirloom Acres Seeds
- Heirloom Seeds
- Heirlooms Evermore Seeds
- High Mowing Organic Seed
- High Mowing Seeds
- Horizon Herbs
- Irish-Eyes
- Johnnys Selected Seeds - 4% Monsanto seed
- Kitchen Garden Seeds
- Lake Valley Seeds
- Livingston Seeds
- Local Harvest
- Mountain Rose Herbs
- Native Seeds
- Natural Gardening Company -% of Monsanto seed
- Nature's Crossroads
- New Hope Seed Company
- Ommas Aarden - Heirloom Seed
- Organica Seed
- Park Seed -% of Monsanto Seed
- Peaceful Valley
- Peaceful Valley Farm Supply
- Pinetree
- Renee's Garden
- Richters Herbs
- Sand Hill Preservation Center
- Seed Saver's Exchange
- Seeds of Change
- Seeds Trust
- Southern Exposure
- Sow True Seed

- Sustainable Seed Company
- Territorial Seed Company - small% of Monsanto seed
- Tomato Fest
- Turtle Tree Seed
- Underwood Garden Seeds
- Uprising Seeds
- Victory Seed Company
- Wild Garden Seed
- Wildseed Farms
- Wood Prairie Farm

[www.seednet.gov.in/Material/Seed_Companies_Pvt_FS_CS.pdfý]

Appendix-7

Major Seed Companies in the World

[http://www.agriquest.info/index.php/seed-companies]

- Ajeet seeds Limited : http://www.ajeetseed.com/
- Produces vegetable seeds like brinjal, chilli, cluster bean, *bhindi*, sponge gourd and bottle gourd.
- Century Seeds: http://www.centuryseeds.com/
- Produces vegetable seeds.
- Eagle Seeds and Biotech Ltd: http://www.eagleseeds.com/
- Produces vegetables seeds.
- Indo-American Hybrid Seeds (India) Pvt. Ltd: http://www.indamseeds.com/
- Produces hybrid vegetable seeds, flower seeds, ornamental plants and biotechnology products.
- JK Agri Genetics Ltd: http://www.jkseeds.net/
- Produces hybrid seeds of tomato, okra and hot pepper.
- Stanagro Seeds: http://www.seedscompany.com/
- Manufacturer and bulk supplier of all kinds of superior quality Indian vegetable seeds. Sungro Seeds: http://www.sungroseeds.com/
- Developed 38 hybrids and 26 superior open pollinated varieties in all major high value vegetable crops like Okra, Bottle gourd, Bitter gourd, Cabbage, Cauliflower, Egg plant, Ridge gourd, Sponge gourd, Tomato, Hot pepper etc.
- Vijrah Exim: http://www.vijrah.com/

- Gold field Biogenetics Pvt. Ltd.- GOFI SEEDS: http://www.gofiseeds.com/
- Syngenta Seeds: http://www.syngenta.com/
- Mahyco Seeds: http://www.mahyco.com/
- Mahyco produces hybrid vegetables.
- Namdhari Seeds: http://www.namdhariseeds.com/
- Breeder, Producer, Distributor for quality vegetable seeds.
- Takii Seeds: http://www.takii.com/
- Takii's breeding program has achieved many "firsts" over the years including the world's first hybrid tomatoes, celery, cabbage, broccoli, turnip, daikon and eggplant.
- Geo Seeds: http://geoseeds.org/
- Producer of select vegetable crop seeds.
- Germin Seeds: http://www.germinseeds.com/
- American Seeds llc: http://americanseedsllc.com/
- Silverhill Seeds: http://www.silverhillseeds.com/
- East African Seeds: http://www.easeed.com/
- Viba Seeds: http://www.vibhseeds.com/
- Nuziveedu Seeds: http://www.nuziveeduseeds.com/
- Nunhems Seeds: http://www.nunhems.com/
- East-West Seeds: http://www.eastwestseed.com/
- Advanta Seeds: http://www.advantaindia.com/
- Vegetable seeds
- Nirmal Seeds: http://www.nirmalseedsindia.com/
- Yaaganti Seeds: http://www.yaagantiseeds.com/
- The company's brand name is "Laxmi Seeds"
- Mansanto Seeds: http://www.monsanto.com/
- Emergegenetic Seeds: http://www.emergegenetics.com/
- Sakata Seeds: http://www.sakata.com/

List of other Seed Companies in the World

- ABT-AgriBioTech Inc. (US).Forage and cool-season turfgrass seed compnay with vertically-integrated platforms in both turfgrass and forage seed.

- Advanta Seeds (US). Has a network of reserch centers, breeding programs, and sales offices. Involved in crops that include corn, canola, sunflowers, and many other types.

- Ag Alumni Seed (US). Provides hybrid popcorn seed, foundation seed stocks, research and testing, private label licensing program, wholesale seed brokerage, and contract seed production.

- AgriGold (US). Develops, produces and markets hybrid seed corn since over half a century.

- AgriPro Seeds (US). Products include corn, alfalfa, wheat, cotton, sunflower, sorghum, and soybean seeds.

- Agstore.net (Ohio, US). Agstore.net supplies farmers in the United States with seed, inoculant, seed treatments, and forage preservatives; features an online catalog of products with easy online ordering and fast shipping.

- Agway (New York, US). Regional enterprises--known as Central Maine, FCI/Vermont, Northeast, Keystone and Southeast --provide animal feeds, seed, fertilizers, crop protectants and other farm supply products and services to farmers.

- Alberta Lea Seed House (US). Primary sales area is the upper midwest of organically raised grains and soybeans. Contact for Spring Field Seed Catalogue.

- Alex Abatti Jr. Farms (US). A group of agriculturally related companies focusing on multiple aspect of farming including forage development, fertilizer development, custom harvesting, and seed cultivation.

- Alpines Unlimited (Canada).Specializes in alpines plant seed sales and information and grows plants from seed for two local nurseries; also offers information on seed germination, growing mediums, and other like topics.

- American Ag-Tec (US).Dedicated to international markets in genetic development and biotechnology, seeds and seed production. Seed research and development also.

- AMPAC Seed Company (US).Wholesale supplier of turfgrass and forage seeds. Related services include contract production, varietal acquisition, and proprietary seed marketing.

- Asgrow Seed Technology (US).Introducing soybean seed with Phytophthora root rot resistance. Offers soybeans with soybean cyst nematode resistance. Offers multiple crop protection and disease resistance stacking in both corn and soybean products.

- B.B.I. - Agricultural Technologies Development And Applications (Israel).B.B.I. is a new Israeli Agriculture technology company specializing in the greenhouse propagation of seedling from tissue-cultures.

- B.V. de ZPC (Netherlands).Seed potatoes and table potatoes.

- Ball Seed Company (US).Wholesale horticultural distributor. Supplies professional growers with seed, plants and greenhouse supplies.

- Barenburg UK (UK).Specializes in standard mixtures of grass seed based not only on STRI, NIAB, SAC and DANI varietal results but also on practical performance.

- Beck's Hybrids (US).Supervision of seed production throughout planting, detasseling and harvesting. Seed conditioning towers on premises.

- BioVision Seed Research Ltd. (Canada).Offers official seed testing services for domestic and import and export of seeds. Seed samples are geminated, seedlings are evaluated, and seed health tested.

- Browning Seed Inc. (US).Developed the first triple cross sorghum-sudangrass hybrid and offers a line of field seeds, including CRP grasses, turf and lawn grass seed, clover, vetc.h, and others.

- Buffel Grass Seed Company (Texas, US). Goal to provide farmers and ranchers with high quality seed and the best services to help maximize their profits along with help in fields such as financial aid, and tech support.
- C & M Seeds (Canada). Offers mixed grain packages that combine complementary varieties based on relative maturities, height, straw strength and other performance traits.
- Cascade Seed Company (US). Produces seed for a global marketplace. Farm hosts a turf and forage grass research program.
- Caudill Seed Co. (US). Produces, processes, and distributes organic sprouts seeds, from a pallet load to a ship load. Can deliver from 100 lbs. to 1000 tons.
- Century Seeds Pvt. Ltd. (India). Varieties of hybrid seeds of tomato, cauliflower, hot peppers, eggplant, radish, and okra.
- China National Tree Seed Corporation (China). Engaged in import and export regarding the tree and shrub breeding materials as seeds, seedlings, grass seeds and flowers.
- Ching Long Seed Co., Ltd (Taiwan). Seeds and breedeing of cauliflowers, broccoli, cabbage, and kale.
- Croplan Genentics (US). Offers a product line that includes corn, soybeans, alfalfa, sunflower, canda, sugarbeet and other crop seeds.
- Cullum Seeds (US). Rice seed which is rouged and certified. Soybeans are tested and germinations are taken.
- De Ruiter Seeds Inc. (US). A subsidiary of DE Ruiter CV Holland, supplies vegetable seed to the greenhouse industry and fresh market processing tomato seed for North America.
- DEF Seeds (US). Offers varieties of corn, soybean, wheat, and alfalfa seeds.
- DEKALB Genetics Corporation (US). Corn, soybeans, sorghum, alfalfa, sunflower, and forage seed products.
- East Texas Seed Company (Texas, US). The company serves the agricultural, turf, and wildlife habitat seed needs. Offers Deer Plot seeds, wildflower seeds, clover seeds, forage seeds and AU Vetc.h seeds.
- Forbes Seed and Grain (US). Produces vegetable seed, both open pollinated and hybrid varieties.

- France Mais Union (France). Production of seed corn, sunflower, cereal, vegtable, fodder plant.

- Gaboriau Technologies (France). Grain, horticulture, and flower seed breeding.

- Garst Corn Hybrid Seeds (Canada). Marketers of seed corn in Canada.

- Goldsmith Seeds Europe BV (Netherlands). Breeding and producing flower seed varieties.

- Graines Gondian (France). Collection of corn and cereals, the production of growing seeds, and the supplying of produce.

- Granite Seed (US). Offers native and domesticated grass, turfgrass, wildflower, and shrub seed. Also carries a full line of erosion control and hydro-seeding products.

- Groupe Cooperatif Occitan (France). Production of seed corn, sunflower, cereal, vegetable, fodder plan.

- Gustafson LLC (US). Researcher, manufacturer and marketer of seed treatment products and related equipment.

- Hazera Quality Seeds (Israel). A breeder, producer, and exporter of vegetable and field crop seeds.

- HBK Seed (US). Production and conditioning of soybeans, rice, wheat, corn, grain sorghum wholesale and retail seed. Also a breeding and research center.

- Heirloom Seeds (US). Over 300 varieties of non-hybrid seeds including vegetable, flower and herb seeds.

- Hettema BV (Netherlands). One of the world's largest in the field of seed potatoes, exporting to over sixty countries.

- Hollarseeds (US). Developer and produer of Cucurbit seed. Over 250 open pollinated and hybrid varieties are produced and sold.

- Howe Seeds (South Dakota, US). Certified and registered seeds for sale, human consumption and the craft industry. Produces flax, barley, wheat, oats, soybeens, ray and other. The majority of the products is in the registered and certified class.

- Independant Grassland Consultants and Seedman (UK). Independant suppliers of grass seed and numerous other products for all customers.

- J.P. Beemsterboer BV (Netherlands). World wide import/export trade company of seed potatoes.

- JH Williams & Sons (Australia). Agricultural seed for sowing temperate species and legumes. Lucernes, subterranean clovers and all pulse crops

- Kraemer & Co Mfg Inc Drying Solutions (California, US). Specializes in drying; years of experience are shown in the application of Column Dryers, Batch Dryers, Box Dryers, and Rotary Dryers which are just a few of the possibilities .

- Limagrain (France). Develops biotechnology, particularly research into geno-mics, in order to improve plants.

- Lur Berri (France). Although seed corn and sweet corn are the mainstay of production, the cooperative has set up many animal branches.

- Maisadour Semences (France). Production of cereals, fodder, crops.

- Mommersteeg International BV (Netherlands). Breeding, production and sales of forage grasses, amenity grasses and catch crops.

- Mommersteeg International Seeds. Seeds of Grasses (Netherlands). Specializes in breeding, seed production, seed processing and sales.

- Momont - La force fertile (France).

- Monsanto Company (NutraSweet, food ingredients). Focuses on three segments: Agricultural (crop protection, biotechnology), Nutrition (NutraSweet, food ingredients) and Pharmaceuticals.

- Mycogen (France). Has developed a line of microbial fatty acid insecticides, fungicides and herbicide products.

- NetSeeds (US). Online seed company that offers the American farmer seed corn, soybeans, alfalfa, sorghum, and grasses. Employment opportunities.

- New Zealand Tree Seeds (New Zealand). Supply a full range of New Zealand native and exotic tree and shrub seeds.

- Nickys Mail Order Seeds (UK). Purveyor of quality seeds worldwide; has an online catalogue of flowers, herbs, wildflowers and mixtures, vegetables, trees, palms, grasses, bulk seed and retail packets .

- Novartis France (France). Dedicated to offering products and services in three areas: Health, Agribusiness, and Consumer health.

- Pacific Seeds (Australia, Thailand). Key products are summer field crops and winter crop seeds. Research to integrate conventional plant breeding techniques with new technology.

- Pannar Seed BV (Netherlands). Deals exclusively in seed for planting and not in grain for milling.

- Pennington Seed (US). Manufacturer of a complete line of consumer lawn and garden products. These include fertilizers, pesticides, grass seed, vegetable seed, soil and forestry products

- Pharmasaat (Artern, Germany). Medicinal and spice plants and seeds. A lot of breeding varieties and species with high yields and high oil contents.

- Premium Seeds Inc. (US). Produces and delivers to the farmer high-yielding hybrid seed-corn

- Quantum Tubers Corporation (US). Rapid-growth biotechnology for potato minitubers. A global company making a quantum leap in seed potato production through rapid-growth biotechnology for pathogen-free, virus-free, potato minitubers. Located in Delavan, Wisconsin.

- Ragt Semences (France). One of Europe's largest seed companies for hybrid corn, sorghum, and forage grasses.

- Research Seeds Inc. (US). Development and commercialization of value-added alfalfa and other forages. Works in the turf market, and the development and distribution of agricultural mircrobial products

- Sakata Seed Europe BV (Netherlands). Flower and vegetable seed.

- Seeds of Rare and Unusual Garden Plants and Flowers (Essex, UK).Seeds of rare and unusual hardy perennial garden plants and flowers, including rock and bog plants, ground cover and climbers all by mail-order to gardeners anywhere in the world. Royal Horticultural Society medal winners.

- Seminis Vegetable Seeds - North & East Europe (Netherlands). Vegetable seed.

- Southern Seed Technology Ltd. (New Zealand). Specializes in nursery services for Northern Hemispere plant and seed companies who wish to multiply spring annuals; crops such as cereals, peas, brassicas, vegetables, flowers and herbs.

- Taylor Garden Seeds (US). Pesticide free seeds. The "Garden Pack" includes 15 select vegetable crop seed packages.

- Teakettle Enterprises, Ltd (Belize). Grower of ornamental plant seed, primarily palms based mainly in Belize.

- Terra Nigra BV (Netherlands). Greenhouse and production laboratories, cultivation of ornamental plants.

- The American Seed Trade Association (US). ASTA is involved in nearly all issues relating to plant germplasm. Site contains events, hot topics, member services, newsletter, and more.

- Thompson Seeds (US). Company does soybean variety research, conditioning, seed treatement-with SMARTCOTE, development and sales.

- TomatoFest Organic Heirlooms (US). Site is a celebration of tomato varieties from around the world; works in creating an extensive marketplace for grown, vine-ripened, heirloom tomatoes and the organic tomato seed business.

- Triumph Seed (US). A diversified seed breeder that markets corn, sunflower, grain, sorghum, forages and soybeans worldwide. Is an independent, privately-owned company.

- Tropical Seeds (Netherlands). A wide variety of green and flowering plant seeds, palm and cycas assortment.

- Vikima Seed (Denmark). Offers custom seed production of important species-hybrids and open pollinated-of vegetables, flowers and herbs.

- West Coast Seeds (British Columbia, Canada). Seed company based on the west coast of Canada. We supply seeds and supplies for farmers and home gardeners. Free catalogue and growing guide.

- Wildseed Farms (US). Offers for sale over 70 species of wildflower seed, and wildflower seed mixes, many of which are grown on their own farm.

- Wyffels Hybrids (US). A regional agricultural seed company providing elite corn hybrids, high oil corn seed blends, and alfalfas to farmer of the US midwest.

[http://www.agriscape.com/companies/seeds/]

Appendix-9

Vegetables with both F_1 hybrid and open-pollinated cultivars showing their adoption trends in the world and their F_1 seed production method.

Vegetables	F_1*	OP*
Asparagus (*Asparagus officinalis*)	Mainly (di)	Old cultivars
Beet and chard (*Beta vulgaris*)	Increasingly (di)	Constant
Bitter gourd (*Momordica charantia*)	Increasingly (h)	Local cultivars
Broccoli (*Brassica oleracea*)	Mainly (si)	Local cultivars
Cabbage (*Brassica oleracea*)	Mainly (si)	Local cultivars
Carrot (Daucus c*arota*)	Increasingly (cms)	Local cultivars
Cauliflower (*Brassica oleracea*)	Mainly (si)	Local cultivars
Celery (*Apium graveolens*)	New (gms)	Mainly
Chinese cabbage (*Brassica rapa*)	Mainly (si)	Local cultivars
Chinese mustard (*Brassica juncea*)	Increasingly (si)	Local cultivars
Cucumber (*Cucumis sativa*)	Mainly (h)	Local cultivars
Eggplant (*Solanum melongena*)	Mainly (h)	Local cultivars
Gourd (*Benincasa hispida*)	Increasingly (h)	Local cultivars
Leek (*Allium porrum*)	New (cms)	Mainly
Luffa (*Luffa angulata* & *L. cylindrica*)	Increasingly (h)	Local cultivars

(contd...)

(contd...)

Vegetables	F_1*	OP*
Melons (*Cucumis melo*)	Mainly (h)	Local cultivars
Okra (*Abelmoschus esculantus*)	Increasingly (h)	Local cultivars
Onion (*Allium cepa*)	Mainly (cms)	Old cultivars
Pakchoi and Petsai (*Brassica rapa*)	Increasingly (si)	Old cultivars
Peppers (*Capsicum annuum*)	Mainly (h)	Local cultivars
Pumpkin (*Cucurbita moschata*)	Increasingly (h)	Old cultivars
Radish (*Raphanus sativus*)	Mainly (si)	Old cultivars
Spinach (*Spinacia oleracea*)	Mainly (di)	Local cultivars
Sweet corn (*Zea mays*)	Mainly (h & cms)	Local cultivars
Tomato (*Lycopersicum esculentum*)	Mainly (h)	Old cultivars
Turnip (*Brassica rapa*)	Mainly (si)	Old cultivars
Watermelon (*Citrullus lanatus*)	Mainly (h)	Old cultivars
Zucchini (*Cucurbita pepo*)	Mainly (h)	Old cultivars

* F_1 – F_1 cultivars; OP – open-pollinated cultivars; (di) – dioecious; (h) – hand-pollinated hybrics; (cms) – cytoplasmic male-sterile system hybrids; (gms) – genetic male-sterile system hybrids; and (si) – self-incompatibility system hybrids.
[*Source:* "Vegetable Hybrid Seed Production," *Seeds: Trade, Production and Technology*, by David Tay.

http://www.seedconsortium.org/PUC/pdf%20files/23-
Vegetable%20Hybrid%20Seed%20Production.pdf]

Number of Plants per acre

Distance (m)	No. of plants (per acre)
1 x 1	4000
2 x 2	1000
3 x 3	444
4 x 4	250
5 x 5	160
6 x 6	111
7 x 7	81
8 x 8	62
9 x 9	43
10 x 10	40

Conversion of pure nutrients to various N, P and K fertilisers

Rate of application (kg/ha)	Urea (46% N)	Super phosphate (18% P)	Muriate of potash (60 % K)
10	22	56	17
20	44	112	34
30	66	168	51
40	88	224	68
50	110	280	85
60	132	336	102
70	154	392	119
80	176	448	136
90	198	504	153

Contd...

Rate of application (kg/ha)	Urea (46% N)	Super phosphate (18% P)	Muriate of potash (60 % K)
100	200	560	170
110	242	616	187
120	264	672	204
130	286	728	221
140	308	784	238
150	330	840	255

Average composition of manures and fertilisers

Manures/Fertilizers	Nutrients (%)		
	N	P	K
Ammonium nitrate	33.5	-	-
Ammonium phosphate	16.0	20.0	-
Ammonium sulphate	20.5	-	-
Ammonium sulphate nitrate	26.0	-	-
Calcium ammonium nitrate	20.5	-	-
Nitrate of soda	16.5	-	-
Urea	46.0	-	-
Single Super Phosphate	-	18.0	-
Double Super Phosphate	-	35.0	-
Triple Super Phosphate	-	45.0	-
Rock phosphate	-	28.4	-
Bone meal	3.5	21.0	-
Muriate of Potash	-	-	50 or 60
Compost	0.5	0.25	0.5
Farm yard manure	0.4	0.3	0.2
Poultry manure	1.2 - 1.5	-	-
Sheep manure	0.8 - 1.6	-	-

Some important varieties

A. Tomato

- Arka Abha (BWR 1)
- Arka Abhijit (BRH 2)
- Arka Ahuti (Sel 11)
- Arka Alok (BER - 5)
- Arka Ashish (IIHR - 674)
- Arka Meghali
- Arka Saurabh (Sel - 4)
- Arka Shreshta
- Arka Vardan (FM hyb -2)
- Arka Vikas (Sel 22)
- Arka Vishal (FM HYB -1)
- Best of All
- Co 1
- Marglobe
- Punjab Chuhra .
- Pusa 120
- Pusa Early Dwarf
- Pusa Ruby
- Rajni
- Rashmi
- Roma
- Rupali
- S-152:
- Sioux

- Vaishali
- Exotic varieties: USA
- *Processing purpose:* Amish Paste, Baylor Paste, Bulgarian Triumph, Carol Chyko's Big Paste, Grandma Mary's, Bellstar, Big Red Paste, Canadian Long Red, Denali, Hungarian Italian, Oroma, Palestinian, Peasant, Polish Paste, Red Sausage, Roma, San Marzano,
- *Table and Processing purpose:* Opalka
- UK
- *Table Purpose:*Gardners delight, Chertia, Evita, Cherry Wonder.
- Russia
- *Processing purpose:*Debarao, Black Plum, Wonder Light
- Italy
- *Processing purpose:* Hogheart, Italian Gold Hybrid, La Rossa VF2, Milano

B. Brinjal

- Annamalai
- APAU Bagyamathi
- APAU Gulabi
- APAU shyamala
- Arka Kusumakar
- Arka Neelkanth
- Arka Nidhi
- Arka Sheel
- Arka shirish
- Aruna
- Azad Kranti
- CO 2
- CO1
- Jamuni Gola
- KKM 1
- Kt4
- MDU 1
- Pant Rituraj
- Pant Samrat
- PKM 1

- PLR 1
- Punjab Barsati
- Punjab Neelum
- Pusa Kranti
- Pusa Purple Cluster
- Pusa Purple Long
- Pusa Purple Round
- F_1 hybrids
- Arka Navneeth
- Azad Hybrid
- Hisar Shyamal (H8)
- MHB - 1
- MHB - 20 (Kalpatharu)
- MHB 9
- Pusa Anmol
- Pusa Hybrid - 5
- Pusa Hybrid - 6

C. Chilli

- Aparna
- Arka Lohit
- Bhagyalakshmi
- CH1
- Co 3
- Co1
- Co2
- Dalchini Suryamukhi
- G3
- Jawahar 218
- K1
- K2
- MDU 1
- Musalwadi
- NP 46A
- Pant C1
- Pant C2

- Punjab Lal
- Pusa Jwala
- Pusa Sadabahar
- Sankeshwar 32
- Sindhur
- Solan
- Ujjwala
- X235

D. Onion

- Agrifound Dark Red
- Agrifound Light Red
- Agrifound Red (Multiplier)
- Agrifound Rose
- Agrifound White
- Akola Safed
- Aprita (RO-59)
- Arka Bindu,
- Arka Kalyan,
- Arka Kirtiman (F_1 hybrid),
- Arka Lalima (F_1 hybrid),
- Arka Niketan,
- Arka Pitambar,
- Arka Pragati,
- Baswant-780,
- Bellary Red (Karnataka)
- Bhima Dark Red,
- Bhima Kiran,
- Bhima Raj,
- Bhima Red,
- Bhima Shakti,
- Bhima Shubra,
- Bhima Shweta,
- Bhima Super,
- Brown Spanish (Long day)
- CO-1(Multiplier)

- CO-2
- CO-3
- CO-4
- Early Grano
- Fursungi Local (Pune)
- Hissar 2
- HOS-1
- K. P. Onion (Andhra Pradesh)
- Kalyanpur Red Round
- **L-28**
- Line-355
- MDU-1
- N-2-4-1
- N-257-9-1
- N-53
- Nasik Red (Nasik)
- Nimar Local (Madhya Pradesh)
- Phule Safed
- Phule Samarth
- Phule Suwarna
- Pillipatti Junagarh (Gujrat)
- Punjab Naroya
- Punjab Naroya
- Punjab Red Round
- Punjab Red Round
- Punjab Selection
- Punjab White
- Punjab White
- Punjab-48
- Pusa Madhavi,
- Pusa Ratnar
- Pusa Red
- Pusa White Flat
- Pusa White Round
- Rajasthan Onion-1

- S-131
- Sukhsagar (West Bengal)
- Telgi Local (Vijapur)
- Udaipur 101
- Udaipur 102
- Udaipur 103
- VL-1 (Long day)
- VL-3 (Long day)

E. **Cabbage**
- Early Golden Acre
- Pride of India
- **Midseason:** Globe
- Resistant Glory and
- Green back
- Badger Shipper
- **Late Seasons:** Success (flat-round)
- Amager Short stemmed
- Pusa Drum head (IARI)
- Sure head (flat round)
- Flat Dutch
- **Cultivars for Storage:** Amager Long stemmed
- Amarger Medium stemmed
- Wisconsin Ball head
- Bodger Ball head
- Holland

F. **Cauliflower**
- Pusa Deepali
- Early Kunwari
- Punjab Giant-26
- Punjab Giant-35
- Pant Shubhra
- Pusa Snowball-1
- Sonwball-16
- Pusa Early Synthetic
- Pant Gobhi-2

- Pant Gobhi-3
- Dania Kalimpong
- Exotic varieties: USA
- Table Purpose: Candid charm, Concert, Cumberland, Fremont, Mariposa, Pathfinder, Ravella, Violet Queen
- G. Radish
- Asiatic Types or tropical types:
- Arka Nishant
- Japanese White
- Kalianpur No. 1
- Nadauni
- Punjab Safed
- Pusa Chetki
- Pusa Desi
- Pusa Reshmi
- European types or temperate types:
- Pusa Himani
- Rapid Red White Tipped
- White Icicle
- H. Cucurbit

Crop	National level	State level
Melon	Kashi Madhu, Pusa Sarbati, Hara Madhu, Pusa Madhuras, MHY-5, , Arka Rajhans, Arka Jeet, Durgapura Madhu, NDM-15, Pusa Rasraj	Punjab Sunehari, Punjab Rasila, Arka Rajhans, Hisar Madhur, RM-43, MHY-3, RM-50, Kashi Madhu, Punjab Hybrid-1, MHY-3, MHL-10, DMH-4
Watermelon	Durgapura Meetha, Sugar Baby, Arka Manik , Arka Jyoti	Durgapura Kesar, Durgapura Lal, RHRWH-12
Bitter gourd	Priya, RHRBG-4-1, KBG-16, PBIG-1, Pusa Hybrid-, NBGH-16	Coimbatore long, Pusa Do Mausmi, Pusa, CO-2, Vishesh, Punjab-14, Kalyanpur Baramasi, CO-1
Pumpkin	CM-14, Pusa Vishwas, Arka Chandan, Arka Suryamukh,i CM-350, NDPK-24	Co-1, Co-2, Narendra Amrit, Kashi Harit, Azad Kaddoo-1
Cucumber	Swarna Ageti, Swarna Sheetal, PCUC-28 UCH-1, Hybrid No.-1, PCUCH-3	Japanese long green, Straight-8, Pusa Uday, Himangi, Swarna Poorna, Sheetal, CO-1, Pusa Sanyog, AAUC-1, AAUC-2

Contd...

Crop	National level	State level
Ridge gourd	Swarna Manjari, PRG-7, Arka Sumeet	Swarna Uphar, Co-1, PKM-1, Arka Sujat, Pusa Nasdar, Punjab Sadabahar, Haritham
Bottle gourd	Pusa Naveen, OBOG-61, NDBG-104, NDBG-132 NDBH-4, PBOG-1, PBOG-2	Arka Bahar, Pusa Sandesh, Pusa Summer Prolific Round, Pusa Summer Prolific long, Punjab Round, Punjab Long, Punjab Komal, CO-1, Narendra Rashmi, Narendra Dharidar, Narendra Shishir , Kashi Ganga, Pusa Manjari, Pusa Hybrid-2, Kashi Bahar
Sponge gourd	Pusa Chikni, CHSG-1, JSGL	Pusa Sneha, PSG-9, Rajendra Nenua-1
Ash gourd	Kashi Ujawal, Pusa Ujawal	Kashi Dhawal

I. Cowpea
- Pusa Barsati
- Pusa Dofasli
- Pusa komal
- Pusa phalguni

J. Indian bean
- Dasara
- Deepali
- Kankan Bushan
- Phule Gauri

K. Palak
- Jobner Green
- Pusa All green
- Pusa harit
- Pusa Jyoti

L. Sweet Potato
- Co-1 (Ib-3)
- Co-2 (Ib-81)
- Co-3
- Cross-4(White)

- H-41 (2)
- H-42 (1)
- Kiran
- Konkal Ashwini (Palgar - 1)
- Konkan Ashwini
- Kufri Garima (Ms/99-1871)
- Kufri Gaurav (JX 576)
- Rajendra Sakarkand-5 (X-5)
- Rajendra Shakarkand-43
- Rajendra Shakarkand-35
- Rnsp-1
- Shalimar Potato-1
- Shalimar Potato-2
- Shree Retna
- Sree Bhadra
- Sree Arun
- Sree Kanaka(X-80/168)
- Sree Varun
- V.L. Sakarkand-6
- Varsha

M. Elephant Foot Yam (Ol)

- Kovvur
- Santragachi

N. Turmeric

- Alleppey Finger
- Amruthapani
- Armoor
- China Scented
- Chinnanadan
- Duggirala
- Erode
- Karhadi
- Local Haldi
- Nizamabad Bulb
- Pattant

- Perianadan
- Rajapore
- Salem Turmeric
- Sangli Turmeric Red Streaked
- Tekurpeta
- Thodopuza
- Waigon
- Wynad

O. Ginger

- Assam
- China
- Himachal
- Himagiri
- IISR Mahima
- IISR Rejatha
- IISR-Varada
- Maran
- Nadia
- Rio-de-Janerio
- Suprabha
- Suravi
- Suruchi
- Cultivars for Andhra Pradesh: Local varieties such as Medak and Tuni
- Cultivars for Arunachal Pradesh: Shillong
- Cultivars for Bihar: Desi and Dorabhanya
- Cultivars for Gujarat: Local types named after respective localities
- Cultivars for Haryana and Punjab: Local types named after respective localities
- Cultivars for Himachal Pradesh: Himachal No. 1, SG 666 (Dhariga local), SG 645 and Narag
- Cultivars for Jammu and Kashmir: Himachal No. 1
- Cultivars for Karnataka: Wynad, Manathodi, Narasapatam Thaiguppan and Karakkal
- Cultivars for Kerala: Kuruppampadi, Wynad local, Valluvanad,

Maran, Nadia, Maran/Lodi, Ernad, Thodupuzha, Rio-de-Janeiro, Jamaica

- Cultivars for Madhya Pradesh: Local ginger types found in Tikkamgarh, Chindwara and Baster districts
- Cultivars for Maharashtra: Local types named after respective localities
- Cultivars for Manipur: Shing type, Thingpuri and Shingtam
- Cultivars for Meghalaya: Nadia, Poona, Rio-de-Janeiro, Wynad, Thingpuri and Maran
- Cultivars for Mizoram: Thingpuri, Maran and Rio-de-Janeiro
- Cultivars for Nagaland: Rio-de-Janeiro
- Cultivars for Orissa: Kuruppampadi, Wynad types, Local types *viz.*, Kuduli, Laxipur, Turia Junagarh, Raikia, Suprapha and Surchi
- Cultivars for Sikkim: Gurubathane, Bhaisey, Nadia, Rio-de-Janeiro and Thingpuri
- Cultivars for Tamil Nadu: Rio-de-Janeiro, Maran, Nadia
- Cultivars for Tripura: Himachal No. 1, Local types
- Cultivars for Uttar Pradesh: Local types named after the localities
- Cultivars for West Bengal: Gorubathan, Sambuk-A, Turuksadar, Malli, Rio-de-Janeiro, Thingpuri, Maran, Tura and Bombay Desi

P. Coriander

- RCr 41
- RCr 20
- RCr 435
- RCr 436
- RCr 446
- GC 1
- GC 2
- Sindhu
- Sadhna
- Swathi
- Co 1
- Co 2
- Co 3
- CS 287
- RD 44 (Rajendra Swathi)

- DH 5

Q. Fenugreek
- Azad Methi-1
- EC- 4911
- Gujrat Methi-1
- Hissar Mukta
- Hissar Sonali
- Kasun
- Kasuri
- Methi No- 14
- Methi No- 47
- ML-150
- Pusa Early Bunching
- RMT-1
- RMT-305

R. Coconut
- Tall Varieties:
- Fiji Tall
- Laccadiv Ordinary (LO)
- Andaman Ordinary(AO)
- Kappadam
- San Ramon
- Philippines
- Spicate
- Pratap
- Philippines Ordinary
- West Coast Tall (WCT)
- East Coast Tall
- Laccadiv Micro
- Semi-Tall Varieties:
- Gangabondam (GB)
- Dwarf Varieties:
- Chowchat Dwarf Orange(COD)
- Chowchat Dwarf Yellow
- Chowghat Dwarf Green

- Malayan Orange Dwarf
- Malayan Yellow Dwarf

Hybrid Varieties
- Anandganga (AO X GB)
- Chandralaksha (LO X COD)
- Chandrasankara (COD X WCT)
- ECT X Gangabondam
- Keraganga (WCT X GB)
- Lakshganga (LO X GB)

S. Arecanut
- Swarnamangala
- Vittal Areca Hybrid - 1 (VTLAH-1)
- South Kanaka
- Thirthahali
- Sree Varjdhan or Rotha
- Mettupalayam
- Kahikuchi
- Mohitnagar
- Mangala (VTL - 3)
- Sumanagala (VTL-11)
- SreeMangala (VTL - 17)

T. Garlic
- GHC-1
- Godawari (Sel. 2)
- HG-17
- JG-99-213 (Gujarat Garlic)
- Vl garlic-1 (VLG-7)
- Vl Lahsun 2 (vgp 5)
- Yamuna   Safed-2
- Yamuna   Safed-3
- Yamuna Safed (G-1)
- Yamuna Safed-4 (g-323)
- Yamuna Safed-5

U. Knolkhol
- Early Purple Vienna

- Early White Vienna
- King of Market
- Large Green
- Palam Tender Knob
- White Vienna

V. Okra

- Arka Abhay (IIHR-4)
- Arka Anamika (IIHR-10)
- Azad Bhindi-1 (Azad Ganga)
- Azad Bhinid-2
- Co.1
- Co-3 (Hybrid-8)
- Co-3 (Hybrid-8) (F)
- Co-3 (Hybrid-8) (M)
- Co-3 (Hybrid-8) (R)
- Gajarat Okra-2
- Gujarat Anand Okra-5 (GAO-5)
- Gujarat Bhinda-1
- Gujarat Okra Hybrid-2
- Hbh-142
- Hisar Naveen (HRB 107-4)
- Hissar Unnat
- Kashi Bhairo (DVR-3)
- Kashi Kranti (VRO-22)
- Kashi Pragati (VRO-6)
- Kashi Satdhari (IIVR-10)
- Kashi Vivhuti (VRO-5)
- Kashilika (IIVR-11)
- M.D.U.I-1
- Parbhani Kranti
- Parkins Long Green
- Phule Kirti Rhroh-4 (Hybrid)
- Phule Utkarshia (GK-IV-3-3-3)
- Punjab Padmini
- Punjab-7

- Pusa Makhmali
- Pusa Sawani
- S-13
- Selection -2
- Shitla Jyoti (DVR-2)
- Shitla Uphar (DVR-1)
- Susthira (AE-286-1)
- Utkal Gaurav
- Varsha Uphar

W. Colocasia

- Bhavapuri (Kcs-2)
- Indira Arvi-1
- Muktakeshi
- Satamukhi
- Sree Karthika (DA-199)
- Sree Kiran (H-13)

X. Garden Pea

- Arka Ajit (FC-1)
- Arkel
- Azad P.I
- Azad P-5
- Azad P-5 (KS-245)
- Bonneville
- Daisy Dwarf
- Early Badger
- Early Giant
- Hara Bona
- Harbhajan
- Hari Chhal
- Jawahar Matar-1
- Jawahar Matar-5
- Jg-63
- Kashi Mukti (VR-22)
- Kashi Nandini (VR-5)
- Kashi Samridhi (VRPMR-11)

- Kashi Shakti (VR-7)
- Kashi Udai (VR-6)
- Lincoln
- Little Marvel
- Madhu
- Meteor
- Narendra Sabzi Matar-4 (NDUP-9)
- Narendra Sabzi Matar-5 (NDUP-250)
- Narendra Sabzi Matar-6 (NDUP-12)
- Pant Uphar (IP-3)
- Perfection New Line
- Punjab-87 (87-1)
- Punjab-88
- Swarna Mukti (CHP-2)
- Vivek Matar 11 (VP 233)
- Vivek Matar-9
- Vl Matar-42

Minimum seed standards

Sl. No	Crop	Submitted sample (in gm)	Working sample	Phy. purity%	Germi- nation %	Moisture%		O.D.V.	
		CS	SS					FS	C
Vegetable crops									
24.	Ash gourd	700	100	70	98	60	7	0	0
25.	Bittergourd V/Hy	1000	250	450	98	60	7	5	10
26.	Bottle gourd	700	100	70	98	60	7	0	0
27.	Ridgegourd Hy/var	1000	150	400	98	60	7	5	10
28.	Snakegourd	1000	250	250	98	60	7	0	0
29.	Watermelon V/Hy	1000	100	250	98	60	7	5	10
30.	Cucumber V/hy	150	100	70	98	60	7	0	0
31.	Pumpkin V/Hy	350	100	180	98	60	7	0	0
Fruit vegetables									
32.	Brinjal V/Hy	150	10	15	98	70	8	0	0
33.	Chillies	150	10	15	98	60	8	0	0
34.	Bhendi	1000	100	140	99	65	10	10	20
35.	Tomato V/Hy	7(hy),70	10	7	98	70	8	0	0
36.	Greens	70	50	7	95	70	8	10	20
Cole crops									
37.	Cabbage	100	10	10	98	70	7	0	0
38.	Caulliflower	100	10	10	98	65	7	0	0
39.	Knol kohl	100	10	10	98	70	7	0	0
Bulb crops									
40.	Onion V/Hy	80	10	80	98	70	8	0	0

Contd...

Appendix-13 Contd...

Sl. No	Crop	Submitted sample (in gm)	Working sample	Phy. purity%	Germi-nation %	Moisture%	O.D.V.	
		CS	SS				FS	C
Vegetable crops								
Root crops								
41.	Carrot V/Hy	30	10	3	95	60	8	5	10
42.	Beet root	500	50	50	96	60	9	0	0
43.	Radish V/Hy	300	50	30	98	70	6	0	0
44.	Turnip V/Hy	70	10	7	98	70	6	0	0
45.	Cluster bean	1000	100	100	98	70	9	10	20
Miscellaneous								
52.	Palak	250	25	25	96	60	9	0	0
53.	Spinach	250	25	25	98	70	–	0	0
54.	Papaya	400	–	40	98	60	12	0	0
55.	Moringa	1000	–	750	96	70	8	0	0
56.	Coriander	400	–	40	95	65	–	0	0

ODV: Other Distinguishing Variety

CS-Certified Sample; SS-Service Sample; FS – Foundation Seed

HY-Hybrid; Var-Variety

Source:

http://www.agritech.tnau.ac.in/seed_certification/seed%20Certification_Quality%20Control.html

Appendix-14

Federal Seed Act Regulations

A. Germination standards for vegetable seeds

Vegetable Seeds	Percent
Artichoke	60
Asparagus	70
Asparagusbean	75
Bean, garden	70
Bean, lima	70
Bean, runner	75
Beet	65
Broadbean	75
Broccoli	75
Brussels sprouts	70
Burdock, great	60
Cabbage	75
Cabbage, tronchuda	70
Cardoon	60
Carrot	55
Cauliflower	75
Celeriac	55
Celery	55
Chard, Swiss	65
Chicory	65
Chinese cabbage	75
Chives	50
Citron	65
Collards	80

Contd...

Contd…

Vegetable Seeds	Percent
Corn, sweet	75
Cornsalad	70
Cowpea	75
Cress, garden	75
Cress, upland	60
Cress, water	40
Cucumber	80
Dandelion	60
Dill	60
Eggplant	60
Endive	70
Kale	75
Kale, Chinese	75
Kale, Siberian	75
Kohlrabi	75
Leek	60
Lettuce	80
Melon	75
Mustard, India	75
Mustard, spinach	75
Okra	50
Onion	70
Onion, Welsh	70
Pak-choi	75
Parsley	60
Parsnip	60
Pea	80
Pepper	55
Pumpkin	75
Radish	75
Rhubarb	60
Rutabaga	75
Sage	60
Salsify	75
Savory, summer	55

Contd...

Vegetable Seeds	Percent
Sorrel	65
Soybean	75
Spinach	60
Spinach, New Zealand	40
Squash	75
Tomato	75
Tomato, husk	50
Turnip	80
Watermelon	70

Source: www.gpo.gov

Appendix-15

A. Desired percentage of moisture in vegetable seeds

Vegetable seeds	Percent
Bean, garden	7.0
Bean, lima	7.0
Beet	7.5
Broccoli	5.0
Brussels sprouts	5.0
Cabbage	5.0
Cabbage, Chinese	5.0
Carrot	7.0
Cauliflower	5.0
Celeriac	7.0
Celery	7.0
Chard, Swiss	7.5
Chives	6.5
Collards	5.0
Corn, sweet	8.0
Cucumber	6.0
Eggplant	6.0
Kale	5.0
Kohlrabi	5.0
Leek	6.5
Lettuce	5.5
Melon	6.0
Mustard, India	5.0
Onion	6.5

Contd...

Contd...

Vegetable seeds	Percent
Onion, Welsh	6.5
Parsley	6.5
Parsnip	6.0
Pea	7.0
Pepper	4.5
Pumpkin	6.0
Radish	5.0
Rutabaga	5.0
Spinach	8.0
Squash	6.0
Tomato	5.5
Turnip	5.0
Watermelon	6.5
All others	6.0

Source: www.gpo.gov

Appendix-16

A. Weight of working sample of seeds (for analysis):

Name of vegetable seed	Minimum weight for purity analysis (grams)	Minimum weight for noxious-weed seed examination (grams)	Approximate number of seeds per gram
Artichoke	100	500	24
Asparagus	100	500	25
Asparagusbean	300	500	8
Bean:			
Garden	500	500	4
Lima	500	500	2
Runner	500	500	1
Beet	50	300	60
Broadbean	500	500	–
Broccoli	10	50	315
Brussels sprouts	10	50	315
Cabbage	10	50	315
Cabbage, Chinese	5	50	635
Cabbage, tronchuda	10	100	–
Carrot	3	50	825
Cauliflower	10	50	315
Celeriac	1	25	2,520
Celery	1	25	2,520
Chard, Swiss	50	300	60
Chicory	3	50	940
Chives	5	50	–
Corn, sweet	500	500	–

Contd...

Contd...

Name of vegetable seed	Minimum weight for purity analysis (grams)	Minimum weight for noxious-weed seed examination (grams)	Approximate number of seeds per gram
Artichoke	100	500	24
Asparagus	100	500	25
Asparagusbean	300	500	8
Bean:			
Garden	500	500	4
Lima	500	500	2
Runner	500	500	1
Beet	50	300	60
Broadbean	500	500	–
Broccoli	10	50	315
Brussels sprouts	10	50	315
Cabbage	10	50	315
Cabbage, Chinese	5	50	635
Cabbage, tronchuda	10	100	–
Carrot	3	50	825
Cauliflower	10	50	315
Celeriac	1	25	2,520
Celery	1	25	2,520
Chard, Swiss	50	300	60
Chicory	3	50	940
Chives	5	50	–
Corn, sweet	500	500	–
Cornsalad:			
Vars. Fullhearted and Dark Green Fullhearted	5	50	–
All other vars	10	50	380
Cowpea	300	500	8
Cress:			
Garden	5	50	425
Upland	2	35	1,160
Water	1	25	5,170
Cucumber	75	500	40

Contd…

Name of vegetable seed	Minimum weight for purity analysis (grams)	Minimum weight for noxious-weed seed examination (grams)	Approximate number of seeds per gram
Dill	3	50	800
Eggplant	10	50	230
Endive	3	50	940
Gherkin, West India	16	160	153
Kale	10	50	315
Kale, Chinese	10	50	–
Kale, Siberian	8	80	325
Kohlrabi	10	50	315
Leek	7	50	395
Lettuce	3	50	890
Melon	50	500	45
Mustard, India	5	50	625
Mustard, spinach	5	50	535
Okra	100	500	19
Onion	7	50	340
Onion, Welsh	10	50	–
Pak-choi	5	50	635
Parsley	5	50	650
Parsnip	5	50	430
Pea	500	500	3
Pepper	15	150	165
Pumpkin	500	500	5
Radish	30	300	75
Rhubarb	50	300	60
Rutabaga	5	50	430
Sage	25	150	120
Salsify	50	300	65
Savory, summer	2	35	1,750
Sorrel	2	35	1,080
Soybean	500	500	6-13
Spinach	25	150	100
Spinach, New Zealand	200	500	13

Contd…

Name of vegetable seed	Minimum weight for purity analysis (grams)	Minimum weight for noxious-weed seed examination (grams)	Approximate number of seeds per gram
Squash	200	500	14
Tomato	5	50	405
Tomato, husk	2	35	1,240
Turnip	5	50	535
Watermelon	200	500	11

Source: www.gpo.gov

Appendix-17

Some Abbreviations

AOSA   : Association of Official Seed Analysts

AOSCA   : Association of Official Seed Certifying Agencies

ISST   : International Society of Seed Technologists

ISTA   : International Seed Testing Association

NSHS   : National Seed Health System

OECD   : Organization of Economic Cooperation and Development

SCST   : Society of Commercial Seed Technologists

[Source: http://www.amseed.org/resources/seed-testing-accreditation-schemes/]

National and State Seed Corporations

- National Seeds Corporation Ltd. (NSC) www.nsc.gov.in
- ANDHRA PRADESH STATE SEED CORPORATION (APSSDC) http://www.apseeds.ap.nic.in
- Assam State Seeds Corporation Ltd., (ASCL)
- Bihar Rajya Beej Nigam Limited (BRBN) http://brbn.bih.nic.in/
- Chhattisgarh Rajya Beej Evam Krishi Vikas Nigam (CGSSC)www.cgbeejnigam.com

- Gujarat State Seeds Corporation Ltd. (GSSC) http://www.gurabini.com
- Haryana Seeds Development Corporation Ltd. (HSDC)
- Jharkhand State Dept of Agriculture (JH)
- Karnataka State Seeds Corporation Ltd. (KSSC)
- Kerala State Seed development Authority (KR)
- Maharashtra State Seed Corporation Ltd. (MSSC) http://www.mahabeej.com
- MP Seeds and Farms Development Corpn. Ltd. (MPS & DC)

[Source: http://seedbank.dacnet.nic.in/Reports/crvSSCReport.aspx]

Appendix-18

SCHEDULE-X

[See Clause 2(xii) and Clause 3(3)]

List of Permit Issuing Authorities for Import of Seeds, Plantsand Plant Products and other articles

[http://www.plantquarantineindia.org/]

S. No (1)	Issuing Authority (2)	Jurisdiction (3)	Authorized to issue permits for (4)
1.	Plant Protection Adviser to the Government of India, Ministry of Agriculture, Directorate of Plant Protection, Quarantine & Storage, NH-IV, Faridabad, Haryana-121001	All notified points of entry	All kinds of plants/plant materials and other items as: soil,peat,insects,microbial cultures, biocontrol agents etc.
2.	Director, National Bureau of Plant Genetic Resources, PUSA Campus, New Delhi-110012	New Delhi	All kinds of import of plant germplasm for public/private sectors Institutions/ in the country.
3.	Joint Director (PP) Plant Quarantine Division, Directorate of Plant Protection, Quarantine and Storage, NH-IV, Faridabad-121001, Haryana.	All notified points of entry	All kinds of plants/plant materials

Contd...

Contd...

S. No (1)	Issuing Authority (2)	Jurisdiction (3)	Authorized to issue permits for (4)
4.	Deputy Director (PP/Ent.), National Plant Quarantine Station, Rangpuri New Delhi-110037.	(i) New Delhi Airport (ii) All Notified points of entry in Northern Zone in the States of Delhi, Haryana, Himachal Pradesh, J&K, Rajasthan, U.P. and Uttaranchal.	Import of all kind of plants/ plant materials for sowing, planting propagation and consumption
5.	DeputyDirector(PP/Ent.), Regional Plant Quarantine Station, Ajnala Road, Near Airforce Station, Raja Sansi Airport, Amritsar-143101	(i) Amritsar Airport (ii) All notified points of entry bordering Pakistan in the States of Punjab & UT Chandigarh	Import of all kind of plants/ plant materials for sowing, planting, propagation and consumption
6.	Deputy Director(PP/Ent.),Regional Plant Quarantine Station ,GST Road, near Trident Hotel, Meenambakam, Chennai-21	(i)ChennaiAirport/Seaport (ii)All notified points of entry in Southern Zone in the States of Andhra Pradesh, Karnataka, Kerala, Tamil Nadu, UTs A&N Islands, Lakshadeep and Pondicherry.	Import of all kind of plants/ plant materials for sowing, planting, propagation and consumption
7.	Deputy Director (PP/Ent.), Regional Plant Quarantine Station, Haji Bunder Road, Sewri, Mumbai-400 014	(i)Mumbai Airport/Seaport (ii)All points of entry notified in Western Zone in the States of Goa, Gujarat, M.P., Chhatisgarh, Maharastra and UT Dadra & Nagar Haveli, Daman & Diu.	i Import of all kind of plants/ plant materials for sowing, planting, propagation and consumption.

Contd...

Contd…

S. No (1)	Issuing Authority (2)	Jurisdiction (3)	Authorized to issue permits for (4)
8.	Deputy Director (PP/Ent.), Regional Plant Quarantine Station,F.B.Block Sector III, Salt Lake City,Kolkata-24	(i) Kolkata Airport/Seaport (ii) All notified points of entry in Eastern Zone in the States of Arunachal Pradesh, Assam, Bihar, Jharkhand, Meghalaya, Manipur, Nagaland, Orissa, Sikkim, Tripura, West Bengal and Mizoram	Import of all kind of plants/ plant materials for sowing, planting, propagation and consumption.
9.	Plant Protection officer (E/PP) Plant Quarantine Station, 24 Paraganas, Bongaon (W.B)	Concerned Port of Entry	Import of Plants and Plant materials for consumption only.
10.	Plant Protection Officer (PP/E), Plant Quarantine Station, 25-A Hariyala Plot, Down Area Bhavnagar-364001/ Kandla	Concerned Port of Entry	Import of Plants and Plant materials for consumption only.
11.	Plant Protection Officer (E/PP), Plant Quarantine Station, Willingdon Island, Cochin.	Concerned Port of Entry	Import of Plants and Plant materials for consumption only.
12.	Plant Protection Officer (E/PP), Plant Quarantine Station, Panitanki- Naxalbari, Rath Khola, P.O. Naxalbari, Distt. Darjeeling (W.B.)	Concerned Port of Entry	Import of Plants and Plant materials for consumption only.
13.	Plant Protection Officer (PP/E), Plant Quarantine Station, Gede Road, Nadiad (W.B.) – 741 503	Concerned Port of Entry	Import of Plants and Plant materials for consumption only.

Contd…

Contd...

S. No (1)	Issuing Authority (2)	Jurisdiction (3)	Authorized to issue permits for (4)
14.	Plant Protection Officer (PP/E) Plant Quarantine Station, 355, Beach Road, Tuticorin – 628 001(T.N.)	Concerned Port of Entry	Import of Plants and Plant materials for consumption only.
15.	Plant Protection Officer (PP/E), PlantQuarantine Station, T.C. No. 28/419, Krishanmurari Road, Kaitha Mukku P.O., Thiruananthpuram – 695 024	Concerned Port of Entry	Import of Plants and Plant materials for consumption only.
16.	Plant Protection Officer (PP/E) Plant Quarantine Station, Harbour, Vishakhapatnam –35	Concerned Port of Entry	Import of Plants and Plant materials for consumption only.
17.	Senior Plant Pathologist Plant Quarantine Station, Cargo Air Terminal Complex, Begampet, Hyderabad –16	Concerned Port of Entry	Import of Plants and Plant materials for consumption only.
18.	Deputy Director (E),Central Integrated Pest Management Centre, Kormangla Road, White field, Bangalore – 70	Concerned Port of Entry	Import of Plants and Plant materials for consumption only.
19.	Plant Protection Officer (E), Central Integrated Pest Management Centre, 16-Professor colony, Bhanwar Kua Main Road, Indore-452001	Concerned Port of Entry	Import of Plants and Plant materials for consumption only.
20.	Plant Protection Officer (E), Central Integrated Pest Management Centre B-16,Mahanagar Extension Lucknow-226020	Concerned Port of Entry	Import of Plants and Plant materials for consumption only.

Contd...

Contd…

S. No (1)	Issuing Authority (2)	Jurisdiction (3)	Authorized to issue permits for (4)
21.	Plant Protection Officer (E), Central Integrated Pest Management Centre, New Punai Chowk Near Dr. Dutta House, Patna – 800 023	Concerned Port of Entry	Import of Plants and Plant materials for consumption only.
22.	Plant Protection Officer (PP/Ent.) CentralIntegrated Pest Management Centre, Mormugao-Harbour, Goa-403 803.	Concerned Port of Entry	Import of Plants and Plant materials for consumption only.

SCHEDULE - IV

[See Clauses 3(2), 10(2) and 11(1) under Plant Quarantine (Regulation of Import into India) Order, 2003]

[http://www.plantquarantineindia.org/]

List of Plants/ Planting Materials and Countries from Where Import is Prohibited along with Justification

Sl. No. (1)	Plant species/ variety (2)	Categories of plant material (3)	Prohibited from the countries (4)	Justification for Prohibition (5)
1.	Banana, Plantain and Abaca (*Musa spp.*)	Rhizomes/Suckers	Central & South America , Hawaii , Philippines and Cameroon	Due to incidence of destructive pests such as Moko wilt (*Burkholderia solanacearum*) race 2 and Cameroon marbling (phytoplasma)
2.	Cassava or tapioca (*Manihot esculenta*)	Seed/Stem cuttings	Africa & South America	Due to incidence of destructive pests such as: Super elongation (*Sphaceloma manihoticola*), Cassava bacterial blight (*Xanthomonas campestris pv. manihotis*) - American strains, Cassava witches' broom (phytoplasma) and several cassava viruses.

Contd...

Contd...

Sl. No. (1)	Plant species/ variety (2)	Categories of plant material (3)	Prohibited from the countries (4)	Justification for Prohibition (5)
3.	Cocoa (*Theobroma cacao*) and plants species belong to Sterculiaceae, Bombacaceae and Tiliaceae.	Fresh beans/ Pods/Bud wood/ GraftsRoot stock/Saplings	West Africa, Tropical America and Sri Lanka .	Due to incidence of destructive pests such as: Swollen shoot virus and related virus strains of cocoa, Witches' broom (*Crinipellis Marasmius*) perniciosa Watery pod rot (*Monilia* (*Moniliopthora*) *roreri*) Mealy pod (*Trachysphaera fructigena*) Mirids (*Sahlbergia singularis & Distantiella theobroma*), Cocoa moth (*Acorocercops cramerella*), Cocoa capsid (*Sahlbergiella theobroma*), Cocoa beetle (*Steirastoma brevi*),Seedling damping-off (*Phytophthora cactorum*), Chestnut downy mildew (*Phytophthora katsurae*) and Black pod of cocoa (*Phytophthora megakarya*).
4.	Cocoyam or Dasheen or Taro (Arvi) (*Colocasia esculenta*) and other edible Aeroids	Plants/Corms/ Cormlets/ Suckers	Cook Islands, Papua New Guinea , Solomon Islands and South Pacific countries	Due to incidence of destructive pests such as Alomae land Bobone (Rhabdo viruses), Dasheen mosaic virus (South Pacific strains) and Bacterial blight (*Xanthomonas campestric pv. dieffenbachiae*).

Contd...

Contd...

Sl. No. (1)	Plant species/ variety (2)	Categories of plant material (3)	Prohibited from the countries (4)	Justification for Prohibition (5)
5.	Coconut (*Cocos nucifera*) and related species of Cocoideae	Seed nuts/ Seedlings/ Pollen/ Tissue cultures etc.	Africa (Cameroon , Ghana , Nigeria , Togo and Tanzania), North America (Florida in USA, Mexico); Central America and Caribbean (Cayman Islands, Bahmas, Cuba, Dominican Republic, Haiti, Jamaica)Philippines and GaumBrazil (Atlantic Coast), Trinidad, Tobago, Greneda, St. Vincent, Barbados, Belize, Honduras, Costa Rica, El Salvador, Panama, Columbia, Venezuela and EcuadorSurinam (Dutch Guyana)Sri Lanka .	Due to incidence of destructive pests such as: Palm lethal yellowing (phytoplasma) and related strains, Cadang cadang & Tinangaja (viroid), Lethal boll rot (*Marasmiellus coco-philus*), Red ring (*Rhadinaphelenchus cocophilus* (*palmarum*), South American Palm weevil (*Rhyncophorus palmarum*), Leaf minor (*Promecotheca cumingi*) and Palm kernel borer (*Pachymerus spp*).
6.	Coffee (*Coffea* spp.) and related species of Rubiaceae	Beans (seeds) / Berries (freshly harvested)/ Grafts/ Bud wood/ Seedlings/Rooted cuttings etc.	Africa andSouth America	Due to incidence of destructive pests such as American leaf spot (*Mycena citricolor, syn. Omphalia flavida*), Coffee berry disease (*Colletotrichum coffeanum var. virulens*), Tracheomycosis (*Gibberella xylariodes, syn Fusarium xylarioids*), Powdery rust (*Hemeleia coffeicola*), Phloem necrosis (Phytomonas leptovasorum) and Coffee viruses (coffee ring spot, leaf rugosity, leaf curl, leaf crinkle and mosaic viruses), Coffee berry borer (*Hypothenemus hampei, Sophronica ventralis*) and Coffee thrips (*Diarthrothrips coffeae*).

Contd...

Contd...

Sl. No. (1)	Plant species/ variety (2)	Categories of plant material (3)	Prohibited from the countries (4)	Justification for Prohibition (5)
7.	Forest plant species:(i) Chestnut (*Castanea* spp.)	(i) Seeds/ Fruits/Grafts and other planting material	North America (USA and Canada)	Due to incidence of destructive pests such as: Chestnut blight or canker (*Cryphonectria* (*Endothia*) *parasitica*)- American strain.
	(ii) Elm (*Ulmus* spp.)	(ii) Plants/ planting material	North America (USA and Canada) and Europe and Russia	Due to incidence of destructive pests such as: Dutch elm disease (*Ceratocystis ulmi*) - American and European strains, Elm mottle virus, Elm bark beetles (*Scolytidae*), Elm phloem necrosis (Phytoplasmas) and White-banded elm leaf hopper (*Scaphoidous luteolus*) -vector of Elm phloem necrosis.
	(iii) Oak (*Quercus* spp.)	(iii) Seeds/ Root grafts	United States of America	Due to incidence of destructive Oak wilt (*Ceratocystis fagacearum*) and Oak bark beetles (*Pseudopityophthorus* spp.)
	(iv) Pine (*Pinus* spp.) and other coniferous species	(iv) (a) Seeds/ Saplings	North America (Canada, USA and Mexico).	Due to incidence of destructive pests such as Pine rusts [Stalactiform blister rust(*Cronartium coleosporioides*), Comandra blister rust (*C. comandrae*), sweet fern blister rust (*C. comptoniae*), Southern fusiform rust (*C. fusiforme*), Western gall rust (*Endocronartium harknessii*), Brown spot needle blight (*Mycosphaerella dearnesii*, syn. *Scirrhia acicola*), Seedling die- back and pitch canker (*Fusarium moniliforme* f.sp. *subglutinans*) and Needle cast (*Lophodermium* spp.)

Contd...

Contd...

Sl. No. (1)	Plant species/ variety (2)	Categories of plant material (3)	Prohibited from the countries (4)	Justification for Prohibition (5)
		(iv) (b) Wood with bark	North America (Canada & USA), Asia (China , Hong Kong , Japan , Korea , Republic of Taiwan)	Due to destructive Pine wood nematode (*Bursaphelenchus xylophilus*)
9.	Oil palm (*Elaeis guineensis*) and related species	Seeds/Pollen/seed sprouts	Philippines and Guam	Due to incidence of Cadang cadang & Tinangaja (*viroid*)
10.	Potato (*Solanum tuberosum*) and other tuber bearing species of Solanaceae	Tubers and other planting material	South America	Due to incidence of destructive pests such as Potato smut [*Thecaphora* (*Angiosorus*) solani],Potato viruses *viz*. Andean potato latent, Andean potato mottle, Arracacha B virus, Potato deforming mosaic, Potato T (capillo virus), Potato yellow dwarf, Potato yellow vein, Potato calico strain of Tobacco ring spot virus and Andean potato weevil (*Premnotrypes* spp.)
11.	Rubber (*Hevea* spp.)	seeds/plants/budwood and any other plant material	Tropical America (Area extending 231/2 degrees North land 231/2 degrees South of the equator (Tropics of Capricorn and Cancer) and includes adjacent islands and longitude 30 degree West land 120 degrees East including part of Mexico, North of the Tropic of Cancer)	Due to incidence of destructive South American Leaf Blight of Rubber (*Microcyclus ulei*)

Contd...

Contd...

Sl. No. (1)	Plant species/ variety (2)	Categories of plant material (3)	Prohibited from the countries (4)	Justification for Prohibition (5)
12.	Sugarcane (*Saccharum* spp.)	Cuttings or sets of planting	Fiji , Papua New Guinea , Australia , Philippines and Indonesia	Due to incidence of destructive Fiji virus
13.	Sweet potato (*Ipomoea* spp.)	Stem (Vine)cuttings rooted or un-rooted/tubers	South Africa , East Africa, New Zealand , Nigeria , USA , Argentina and Israel .	Due to incidence of destructive pests such as: Scab (*Elsinoe batatas*), Scurf (*Moniliochaetes infuscans*), Foot rot (*Plenodomus destruens*), Soil rot (*Streptomyces ipomoeae*), Bacteria wilt (*Pseudomonas batatae*), Sweet potato viruses *viz.* Russet crack; feathery mottle; internal cork; chlorotic leaf spot; vein mosaic; mild mottle and yellow dwarf, vein clearing; chlorotic stunt; Sheffied's virus A and B etc., Sweet potato witches' broom (*phytoplasmas*) and seed bruchid (*Mimosestes mimosae*)
14	Yam (*Dioscorea* spp.)	Tubers for planting or propagation	West Africa and Caribbean region	Due to incidence of destructive Yam mosaic virus/ green banding virus

Import of plant and plant materials – Flow chart

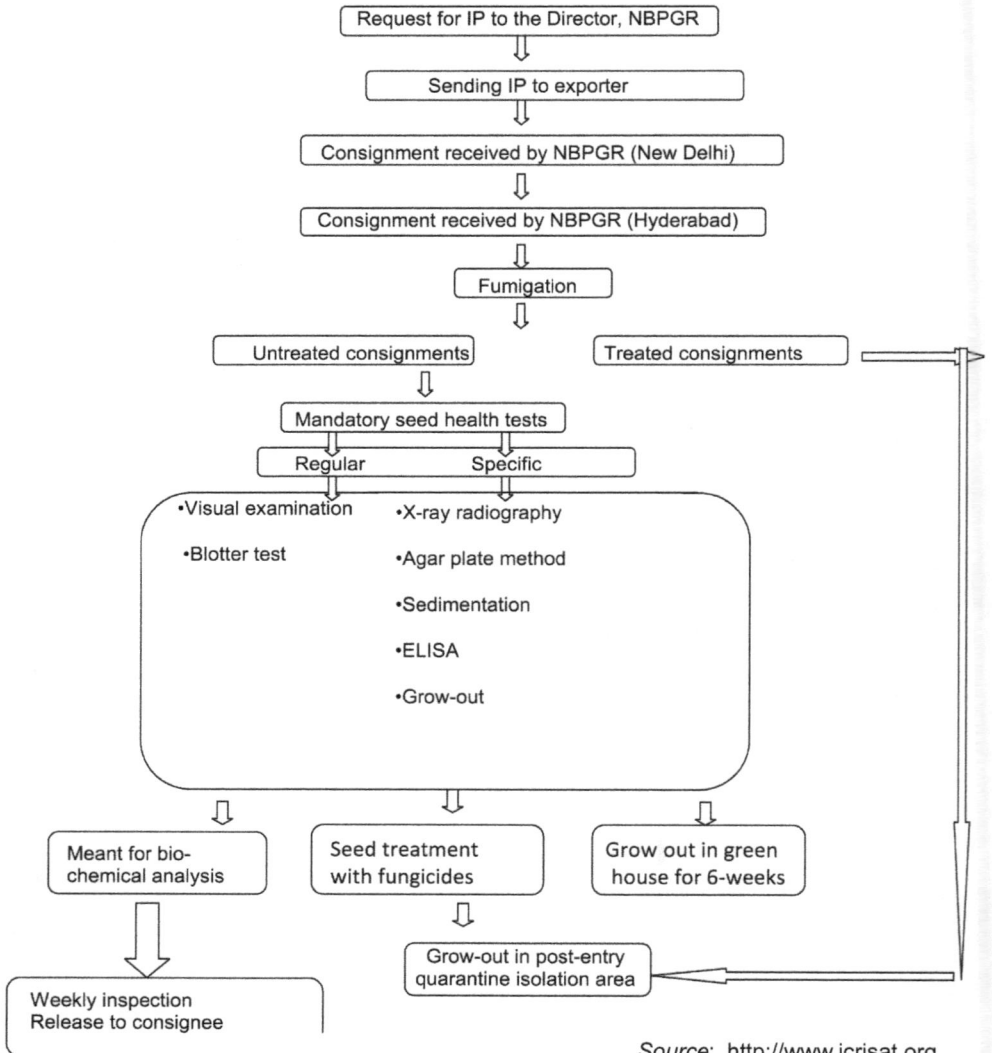

```
Request for IP to the Director, NBPGR
                  ⇓
        Sending IP to exporter
                  ⇓
  Consignment received by NBPGR (New Delhi)
                  ⇓
  Consignment received by NBPGR (Hyderabad)
                  ⇓
            Fumigation
                  ⇓
  Untreated consignments          Treated consignments
                  ⇓
      Mandatory seed health tests
        Regular        Specific
```

Regular	Specific
•Visual examination	•X-ray radiography
•Blotter test	•Agar plate method
	•Sedimentation
	•ELISA
	•Grow-out

```
Meant for bio-        Seed treatment       Grow out in green
chemical analysis     with fungicides      house for 6-weeks
       ⇓                    ⇓
                    Grow-out in post-entry
                    quarantine isolation area
Weekly inspection
Release to consignee
```

Source: http://www.icrisat.org

Appendix-21

SCHEDULE-I

[See clauses 2 (xxi), 3 (13) and 3 (14)

[http://dbtbiosafety.nic.in]

Points of Entry for Import of plants/plant materials and other Articles

Seaports	Airports	Land Frontier Stations
1. Alleppey (Kerala)*	1. Amritsar (Punjab)	1. Agartala (Tripura)
2. Bhavnagar (Gujarat)	2. Bangalore (Karnataka)	2. Amritsar Rly. Stn. (Punjab)
3. Kolkata (West Bengal)	3. Kolkata (WB)	3. Attari Rly. Stn. (Punjab)
4. Calicut (Kerala)	4. Chennai (TN)	4. Attari Wagha Border Chk post (Punjab)
5. Chennai (Tamil Nadu)	5. Cochin (Kerala)	5. Bongaon (WB)
6. Cochin (Kerala)	6. Hyderabad (AP)	6. Gede Road Rly. Stn. (WB)
7. Cuddalore (TN)*	7. Mumbai (Maharashtra)	7. Karimganj (Assam)
8. Goa* (Goa)	8. Palam (New Delhi)	8. Kishanganj (WB)
9. Gopalpur (Orissa)*	9. Patna (Bihar)	9. Moreh (Manipur)
10. Haldia (WB)*	10. Tiruchirapalli (TN)	10. Panitanki (WB)
11. Jamnagar (Gujarat)*	11. Trivandrum (Kerala)	11. Raxual (Bihar)
12. Beypore (Kerala)*	12. Varanasi (UP)	12. Rupehiya (UP)
13. Kakinada (AP)		13. Sonauli (UP)
14. Kandla (Gujarat)		
15. Karwar (Karnataka)*		
16. Krishnapatnam (AP)*		
17. Machlipatnam (AP)*		
18. Mandvi (Gujarat)*		
19. Mangalore (Karnataka)*		

Contd...

Contd..

Seaports	Airports	Land Frontier Stations
1. Mumbai (Maharashtra)		
2. Mundra (Gujarat)*		
3. Nagapatnam (TN)*		
4. Nova Shiva (Maharashtra)		
5. Navlakhi (Gujarat)*		
6. Okha (Gujarat)*		
7. Paradeep (Orissa)*		
8. Pondicherry*		
9. Porbander (Gujarat)*		
10. Rameshwram ((TN)		
11. Tiruvananthapuram (Kerala)		
12. Tuticorin (TN)		
13. Veraval (Gujarat)*		
14. Visakhapatnam (AP)		
15. Vizhinjam (Kerala)*		

* For import of food grains and timber only.

Appendix-22

FORM-A

[Clause 3(2)(i)][http://agricoop.nic.in]

Application for Permit to Import Seeds/Fruits/Plants for Consumption

To

The Competent Authority

The undersigned hereby applies for a permit authorizing the import seeds/fruits for consumption as per details given below:

(Please write/type in Block Letters)

1. Name and exact description of seeds/fruits/plants to be imported:

2. Description of the Consignment and Quantity:

3. Name and address of the consignor:

4. Name and address of the importer:

5. Country and locality in which grown or produced:

6. Foreign port of shipment:

7. Approximate date of arrival of the consignment in India:

8. Name of *Air/Sea Port/Land Customs Station of entry in India:

I undertake to produce an official Phytosanitory Certificate with additional declaration, if any as specified in the permit. I also undertake to pay to the Plant Protection Adviser or any officer duly authorized by him, the prescribed fees to meet the cost of inspection, fumigation, disinfestation and disinfection of the consignment referred to above.

Signature of the importer or his authorised agent.

Place:

Name and postal address of the importer or his authorized agent.

Date:

*Strike out whichever is not applicable

FORM –B

[See Clause 3, condition (2)]

__Application For Permit To Import Seeds And Plants For Sowing/Planting__

To,

The Competent Authority,

The undersigned hereby applies for a permit authorizing the import of Seeds/Plants as per details given below: -

1. Name and address of importers: (Please type/write in BLOCK LETTERS)

2.

Sl. No.	Exact description of seeds/plant to be imported (state commercial and botanical name)	Name of Hybrid/variety	Quantity	
			No. of packages	Total weight or total No. of propagating material
1	2	3	4(a)	4(b)

3. *Catalogue of seeds producer establishing identity of the seed/planting material to be imported:

4. Name and address of the producing company:

5. Name and address of consignor:

6. Country and locality in which seed planting material

7. Foreign port of shipment:

8. Approximate date of arrival of consignment in India and name of **Air/Seaport/Land Custom Station:

9. Number and date of registration certificate from National Seeds Corporation/State Director of Agriculture/Horti-Culture/Central/ State Govt. Authorities (alongwith a photo copy):

10. Number and date of import licence for commercial import of seeds of coarse cereals, oilseeds and pulses (photo copy to be attached:

11. ONLY FOR IMPORTS FOR WHICH POST ENTRY QUARANTINE/INSPECTION IS PRESCRIBED

    a. Exact locality and its postal address where imported

    b. Seeds/plants will be grown:

    c. Names, postal address of Designated Inspection

    Agency (DIA) under whose supervision imported   Seeds/plant will be grown:

4. DECLARATION

I declare that the information furnished is correct to the best of my knowledge and belief.

I undertake to produce an official Phytosanitary Certificate with additional declarations, if any,as specified in the permit

I also undertake to pay to the Plant Protection Adviser or the officer duly authorized by him, the prescribed fees to meet the cost of inspection, fumigation, disinfestation and disinfection of the consignment referred to above.

Place:

                    Signature of the Importer or his authorized agent.

Date:

- Photo copy of cover page and the relevant portion, if original catalogue cannot be furnished/photo copies of documents establishing identity of the seeds/planting materials.

@ Only for Food Processing Industries.
**Strike out whichever is not applicable

FORM – C
[Clause 3(3)]
(National Emblem)
Government of India
MINISTRY OF AGRICULTURE
(Department of Agriculture & Cooperation)
Directorate of Plant Protection, Quarantine & Storage
N.H. IV., Faridabad

Permit For Import Of Fruits/Seeds/Plants For Consumption

Permit No._____
Date_____ Valid upto _____

1.   Permission is hereby granted _____

_____

(Name and address of the importer or his authorized agent)

to import by air/sea/land the plants/seeds fruits herein specified grown or produced in _____ from _____
(Name and address of the consignor) through air/sea port /land custom station

_____

(Name of Port/Station)

As per following details:
1. Name and exact description of seeds/fruits/plants/to be imported
2. Description of the Consignment and Quantity
3. Country and locality in which grown or produced.
4. Foreign Port of shipment

5. Specific purpose of import

 1. The consignment should be :

  i. Accompanied by an official Phytosanitary Certificate issued by the authorized officer of the country of origin (i.e.*

  ii. The official Phytosanitary Certificate shall also contain the following additional declarations: -

Note : (1) The importer or his authorized agent shall produce this permit for inspection by the Plant Protection Adviser or an officer authorized by him at the time of arrival of the consignment at the land customs station or port of entry.

 1. THE IMPORTER SHALL INTIMATE IMMEDIATELY TO THE PERMIT ISSUING AUTHORITY OF ANY CHANGE OF ADDRESS

 Here specify the country of origin

The gazette of India; extraordinary [part ii-sec-3 (ii)]

FORM –I

[Clause 3(4)]

FACE OF TAG

This package contains perishable plants/seeds

Rush and Deliver to:

     The Officer In-charge

     Plant Quarantine and Fumigation Station,

     _____Airport/seaport/Land Customs Station

                    Signature of competent Authority

REVERSE OF TAG

Permit No._____Valid upto_____

Directions for sending plants/seeds

Under this tag only materials covered by

Permit the number of which it bears should be booked

Any other material may be confiscated

Place inside the package the importer's name and address, the invoice, and an official Phytosanitary Certificate issued by authorized officers of your country. In case of import by sea rush all documents to consignee by air.

Attach Tag securely to consignment.

SCHEDULE –I
[CLAUSE 2(1)]
OFFICIAL PHYTOSANITARY CERTIFICATE

Plant Protection Organization of _____ No._____
(Name of the country)

To

The Plant Protection Adviser to the Govt. of India
Dte. of Plant Protection, Quarantine and Storage
N.H.IV.Faridabad-121001
(India)

DESCRIPTION OF CONSIGNMENT

Name and address of the exporter _____

Declared name and address of consignee _____

Number and description of packages _____

Distinguishing marks: _____

Place of origin _____

Declared means of conveyance _____

Declared port of entry/land custom station _____

Name of the produce and the quantity declared : _____

Commercial and Botanical names of the plants/seeds/ fruits_____

This is to certify that the plants or plant products described at above have been inspected and free from quarantine pests and substantially free from other injurious pests; and that they are considered to Phytoanitary regulations of the importing country.

DISINFESTATION AND/OR DISINFECTION TREATMENT

Date_____ Treatment_____

Chemical (active ingredient)_____Duration and temperature_____

Concentration_____Additional information _____

Additional declaration

(Stamp of Organization)

Place of Issue: _____

Name of the authorized officer_____

Signature

Date _____

No financial liability with respect to this certificate shall attach to _____(Name of the Plant Protection Organization) _____or any of its officers or representative. _____.

*Option clause.